INTERNATIONAL CENTRE FOR MECHANICAL SCIENCES

COURSES AND LECTURES - No. 313

GEOMETRIES, CODES AND CRYPTOGRAPHY

EDITED BY

G. LONGO
UNIVERSITY OF TRIESTE

M. MARCHI
UNIVERSITY OF UDINE

A. SGARRO
UNIVERSITY OF UDINE

SPRINGER-VERLAG WIEN GMBH

Le spese di stampa di questo volume sono in parte coperte da
contributi del Consiglio Nazionale delle Ricerche.

This volume contains 36 illustrations.

Originally published by Springer Verlag Wien-New York in 1990

In order to make this volume available as economically and as rapidly as possible the authors' typescripts have been reproduced in their original forms. This method unfortunately has its typographical limitations but it is hoped that they in no way distract the reader.

ISBN 978-3-211-82205-0 **ISBN 978-3-7091-2838-1 (eBook)**
DOI 10.1007/978-3-7091-2838-1

PREFACE

Geometry is the art not to make calculations.

The International Centre for Mechanical Sciences has a long-standing tradition of advanced schools in coding theory and information theory, started as early as 1970 (cf the list of volumes in this series at the end of this book). In 1983 a rather exciting event took place, namely an advanced school on the new subject of cryptology (queer indeed, that the the century-long art of secret writing should be called new!). The 1983 meeting (cf volume n. 279 in this series) has by now entered the history of contemporary European cryptologic research, preceded as it is only by one similar event in Burg Feuerstein, Germany, in 1982, and followed by the 1984 workshop at the Sorbonne of Paris, which was the first of this sort to bear the by now well-established name of Eurocrypt.
So, in a way, the Udine meeting was Eurocrypt number zero, second in a list whose origin is curiously set at n= -1.

After this, CISM kept secretive for several years on the subject of coding, be it source coding, channel coding or ciphering. In 1989 the second editor, M. Marchi, who is a "pure" geometer, having heard about the spectacular success that finite (pure!) geometries were experiencing in the domain of authentication schemes (a very matter-of-fact subject nowadays, as it has so much to do with the proper handling of our credit cards), decided to appease his curiosity and contacted the third editor, A. Sgarro, an information theorist active in Shannon-theoretic cryptology. From their conversations the idea of this school originated; the first editor, G. Longo, himself an information theorist but also a writer, added his experience to their enthusiasm. The result was a delightful week during which students and researchers alike were given the opportunity to meet, to exchange ideas, to get along in their field, and, why not!, to make friends. About fifty people participated, coming from fifteen different countries. Geometers and coding theorists (not necessarily distinct persons) worked shoulder to shoulder in the historical Palazzo del Torso, the beutiful 16th-century site of CISM; we are confident that these contacts helped to blur the largely artificial borders between "pure" and "applied" mathematics.

While classes took place in the Palazzo del Torso, conversation groups could spread all over the charming town of Udine, which is a fertile crossroads of European cultures, situated as it is at the intersection of Latin, Germanic and Slav areas. We are particularly proud of this fact in this 1989 of wonders, which, we firmly hope, will open an age of thriving for what was often and rather bitterly called the old continent.

The editors: Giuseppe Longo, Mario Marchi, Andrea Sgarro

CONTENTS

PART ONE

Geometries and Codes

LECTURES ON GALOIS GEOMETRIES AND STEINER SYSTEMS

G. Tallini
Università La Sapienza, Roma, Italy

ABSTRACT

The paper consists of four lectures held at "Centre International des Sciences Meccaniques" of Udine (Italy), June 1989. Their content is the following. General concepts on Galois geometry and Steiner systems. The theory of h-sets in Steiner systems, with particular attention to Galois spaces. The theory of blocking sets, the even and odd type sets in a Steiner system. Applications to linear error correcting codes.

1. GALOIS SPACES

Let p be a prime, Z_p the field of the residue classes mod p, g(x) a polynomial with coefficients in Z_p, irreducible in Z_p and of degree h. We call *Galois field* of order $q = p^h$ the field $GF(q) = Z_p[x]/(g(x))$, algebraic extension of Z_p by the polynomial g(x). We prove that *every finite field has order $q = p^h$ (p a prime) and it is isomorphic to a Galois field.* Moreover, *for every prime p and integer h a Galois field of order $q = p^h$ exists and it is unique up to an isomorphism.*

For any integer $r \geq 2$ we denote by PG(r,q) *the projective space* of dimension r over GF(q). The points of PG(r,q) are therefore the ordered (r+1)-tuples of not all zero elements of GF(q), determined up to a non zero multiplicative factor in GF(q). A subspace S_d of dimension d in PG(r,q) is the set of points whose coordinates satisfy a system of r-d linear homogeneous independent equations.

We set

$$\theta_d = \sum_{i=0}^{d} q^i . \tag{1.1}$$

We prove:

I. - *In PG(r,q) a subspace S_d, $1 \leq d \leq r$, has θ_d points:*

$$|S_d| = \theta_d \tag{1.2}$$

In particular:

$$|PG(r,q)| = \theta_r \tag{1.3}$$

II. - *The number of hyperplanes, S_{r-1}, of PG(r,q) is θ_r. It fol-*

lows that the number of S_{d-1}*'s belonging to a* S_d *of* $PG(r,q)$ *is* θ_d.

III. - *In* $PG(r,q)$ *the number of hyperplanes through a* S_{r-d-1} *is* θ_d.

IV. - *We denote by* $\gamma_{r,d,q}$, $1 \le d \le r-1$, *the number of subspaces* S_d *of* $PG(r,q)$. *It is:*

$$\gamma_{r,d,q} = \prod_{i=0}^{d} (\theta_{r-i}/\theta_{d-i}) \ . \tag{1.4}$$

V. - *In* $PG(r,q)$ *the number of* S_d*'s through a given* S_m, $m \ge 0$, *is* $\gamma_{r-m-1,d-m-1,q}$.

2. STEINER SYSTEMS S(2,k,v)

A *Steiner system* $S(2,k,v)$ (or a *2-(v,k,1) design*) is a pair (S,L), where S is a set whose elements we call points and L is a family of parts of S whose elements we call lines such that:

$$|S| = v; \qquad \forall \ell \in L \Rightarrow |\ell| = k \ . \tag{2.1}$$

through two distinct points there is a unique line:

$$\forall P,Q \in S, P \ne Q \Rightarrow \exists! \ \ell \in L: \ P,Q \in \ell \tag{2.2}$$

For example PG(r,q) with respect to its lines is a $S(2,q+1,\theta_r = \sum_{i=0}^{r} q^i)$.

We denote by F_p the set of lines of S(2,k,v) through a point P of S. It is:

$$\forall P \in S, \quad |F_p| = (v-1)/(k-1) \tag{2.3}$$

Counting in two different ways the pairs (P,ℓ) where $P \in S$, $\ell \in L$, $P \in \ell$, we have $v.|F_p| = |L|k$, then (see (2.3)):

$$|L| = v(v-1)/k(k-1) \; . \tag{2.4}$$

Therefore a necessary condition in order a $S(2,k,v)$ exists, is the following:

$$r = (v-1)/(k-1) \; , \quad b = v(v-1)/k(k-1) \tag{2.5}$$

are both integers.

If $P \in S$ and $\ell \in L$, $P \notin \ell$, we denote by u the number of lines through P not meeting the line ℓ (*parallel* to ℓ). It is:

$$u = r - k \tag{2.6}$$

We have:

$$0 \leq u = r - k = (v-1)/(k-1) - k \Longrightarrow v \geq k(k-1) + 1 = k^2 - k + 1$$

It is:

*$u = 0 \iff v = k^2 - k + 1 \iff [\forall \ell, \ell' \in L, \ell \neq \ell' \Rightarrow |\ell \cap \ell'| = 1] \iff$ (2.7)
$\iff S(2,k,k^2 - k + 1)$ is a projective plane of order $q = k-1$,
$S(2,k,k^2-k+1) = S(2,q+1,q^2+q+1) \iff |L| = |S| = q^2 + q + 1$;
$\forall P \in S$, $\forall \ell \in L$, $|\ell| = q+1 = |F_p|$, for example $PG(2,q)$.*

*$u = 1 \iff v = k^2 \iff [\forall P \in S, \forall \ell \in L, \exists! \ell' \in L: P \in \ell', \ell' \cap \ell = \emptyset$ (2.8)
(ℓ' parallel to ℓ)] $\iff S(2,k,k^2)$ is an affine plane of*

order k, for example AG(2,q), k = q.

We call *subspace* of $S(2,k,v)$ a subset T of S such that:

$$P,Q \in T \implies \text{the line } PQ \subseteq T. \tag{2.9}$$

Obviously $\emptyset$, the points, the lines, S, are subspaces. Moreover every subspace T ($|T| \geq 2$) is a $S(2,k,v_T)$, with $v_T \leq v$ and, if $T \neq S$, $v_T = |T| \leq r$.

The family Σ of subspaces of (S,L) is a *closure system*, that is

$$\begin{cases} S \in \Sigma \\ \{T_i\}_{i \in I},\ T_i \in \Sigma \implies \bigcap_{i \in I} T_i \in \Sigma \end{cases} \tag{2.10}$$

If $X \subseteq S$, the *closure*, $\bar{X}$, of X is the intersection of all subspaces containing X, that is the *minimal* subspace containing X.

A set I of points of S is called *independent* if:

$$x \in I \implies x \notin \overline{(I - \{x\})}. \tag{2.11}$$

A *base* of (S,L) is a *maximal independent* set of S.

(S,L) is a *matroid* if:

$$\forall X \subseteq S,\ \forall I,\ I' \text{ maximal independent contained in } X \implies |I| = |I'|. \tag{2.12}$$

If (S,L) is a matroid we define:

$$\text{rank } X = |I|,\ I \text{ maximal independent of } X, \tag{2.13}$$
$$\dim(S,L) = \text{rank } S - 1.$$

If (S,L) is a Galois space $PG(r,q)$, the subspaces of (S,L) are the subspaces of $PG(r,q)$. The closure of $X \subseteq S$ is its linear closure in $PG(r,q)$. Moreover $PG(r,q)$ is a matroid and $\dim(S,L) = r$.

3. GENERAL CONCEPTS ABOUT h-SETS OF A S(2,k,v)

A h-set of $S(2,k,v) = (S,L)$ is a subset H consisting of h points of S. We classify such sets with respect to their behaviour with the lines of (S,L).

We define *s-index character of H* the number t_s of lines meeting H in s points, $0 \leq s \leq k$. It is:

t_0 = Number of external lines to H,

t_1 = Number of tangent lines to H,

t_2 = Number of 2-secant lines to H,

t_s = Number of s-secant lines to H,

t_k = Number of lines belonging to H.

We prove:

I. - *The characters of a h-set of $S(2,k,v)$ satisfy the following equations:*

$$\begin{cases} \sum_{s=0}^{k} t_s = b = v(v-1)/k(k-1), \\ \sum_{s=0}^{k} s\, t_s = h\, r = h(v-1)/(k-1), \\ \sum_{s=0}^{k} s(s-1)t_s = h(h-1). \end{cases} \tag{3.1}$$

By $(3.1)_2$ and $(3.1)_3$ we have:

$$\sum_{s=0}^{k} s^2 t_s = h(h-1+r) = h[h-1 + (v-1)/(k-1)]. \tag{3.2}$$

We say that H is an *m-character set*, if exactly m characters are different from zero. Let $(n_1, n_2, \ldots, n_r)$ be integers such that $0 \leq n_1 < < n_2 < \ldots < n_r \leq k$. We say that H is of *class* $[n_1, n_2, \ldots, n_r]$ if $t_n = 0$, $\forall\, n \neq n_1, n_2, \ldots, n_r$. We say that H is of *type* $(n_1, n_2, \ldots, n_m)$ if it is of class $[n_1, n_2, \ldots, n_m]$ and $t_{n_i} \neq 0$ $(i=1,2,\ldots,m)$.

The study of the h-sets of S(2,k,v) may be done by increasing the number of non-zero characters. We have:

II. - *If H has a unique character, either $H = \emptyset$ or $H = S$, that is non-trivial one-character sets don't exixt.*

Proof. If $H \neq \emptyset$, $H \neq S$, a point $P \in H$ and a point $Q \in S - H$ exist. Let be $n = |H \cap \ell|$, $\ell \in L$. If we range the points of H, different from P, on the r lines through P, we have $|H| = r(n-1)+1$. For Q we have similarly $|H| = r.n$, whence $rn = r(n-1)+1$, a contradiction (since $r \geq k > 1$).

Assume H has two characters, so that it is of type (m,n), with $0 \leq m < n \leq k$. By (3.1) we have:

$$\begin{cases} H \text{ of type } (m,n) \implies t_m = (nb-hr)/(n-m), \quad t_n = (hr-mb)/(n-m) \\ h^2 - h[1 + (n+m-1)r] + m\,n\,b = 0 \\ \Delta = [1 + (n+m-1)r]^2 - 4\,m\,n\,b \quad \text{is a square.} \end{cases} \tag{3.3}$$

Moreover we prove:

III. - *Sets of type (0,k) don't exist. The sets of type (1,k) are the subspaces of S(2,k,v) such that every line meets them (we call such subspaces, Π, primes of S(2,k,v) and we have |Π| = r). The sets H of type (0,n), n <k, are such that:*

$$\begin{cases} |H| = h = (n-1)r + 1 , \\ P \notin H, \ \sigma_P = \# n\text{-secant lines through } P = r - (r-1)/n, \\ n \text{ divides } r-1. \end{cases} \qquad (3.4)$$

The sets H' of type (m,k) are the complements of those of type (0,n = = k-m), whence:

$$\begin{cases} |H'| = v - (k-m-1)r - 1, \\ k - m \text{ divides } r - 1. \end{cases} \qquad (3.4')$$

IV. - *Every set of class [0,1,k] of S(2,k,v) is a subspace and conversely.*

We call *(h;m,n)-set* of S(2,k,v) a h-set H (≠ Ø,S) such that:

$$m = \min_{\substack{\ell \in L \\ \ell \cap H \neq \emptyset}} |\ell \cap H| , \qquad n = \max_{\ell \in L} |\ell \cap H| . \qquad (3.5)$$

We have:

$$h \leq m + (r-1)n , \qquad (3.6)$$

$$t_0 = 0,\ n < k \implies h \geq n + (r-1)m , \qquad (3.7)$$

Let M,N be two integers such that $1 \leq M \leq m$, $n \leq N$. By (3.1), (3.2) for any (h;m,n)-set H of S(2,k,v) we have:

$$0 \leq \sum_{M}^{N} (N-s)(s-M)t_s = -MN \sum_{M}^{N} t_s + (M+N) \sum_{M}^{N} s\, t_s - \sum_{M}^{N} s^2 t_s =$$

$$= - MN(b-t_0) + (M+N)hr - h(h-1+r) =$$

$$= -h^2 + h[r(M+N-1) + 1] - MN(b-t_0)$$

that is:

$$0 \leq \sum_{M}^{N} (N-s)(s-M)t_s = -h^2 + h[r(M+N-1) + 1] - MN(b-t_0) \tag{3.8}$$

It follows:

$$\begin{cases} MNt_0 \geq h^2 - h[r(M+N-1) + 1] + MNb \\ = \iff N = n,\ M = m,\ H \text{ is of class } [0,m,n], \end{cases} \tag{3.9}$$

whence:

$$t_0 = 0 \implies \begin{cases} h^2 - h[r(M+N-1) + 1] + MNb \leq 0 \\ \text{the equality holds} \Leftrightarrow M=m, N=n,\ H \text{ is of type } (m,n) \end{cases} \tag{3.10}$$

$$t_0 = 0 \implies \begin{cases} \Delta = [r(M+N-1) + 1]^2 - 4MNb \geq 0 \\ \Delta = 0 \iff M=m, N=n, H \text{ is of type } (m,n), h=[r(m+n-1)+1]/2 \end{cases} \tag{3.11}$$

By (3.11), (3.10) it follows that:

V. - *In $S(2,k,v)$ sets of class $[1,2,3]$ don't exist, if $4v > k(3r+2)$. In particular in $S(2,q+1,q^2+q+1)$ and in $S(2,q,q^2)$ sets of class $[1,2,3]$ don't exist if $q \geq 5$.*

4. INTERSECTION SETS AND BLOCKING SETS IN S(2,k,v)

An *intersection set* of S(2,k,v) is a set H meeting every line in at least one point, that is such that $t_0 = 0$. It is:

$$H \textit{ intersection set } \Longrightarrow \begin{cases} |H| \geq r\,, \\ |H| = r \iff H \textit{ is of type } (1,k) \iff \\ \iff H \textit{ is a prime} \end{cases} \tag{4.1}$$

A *blocking set* in $S(2,k,v)$ is an intersection set not containing lines, that is such that $t_0 = t_k = 0$. Obviously, *if* B *is a blocking set, also* $S - B$ *it is.* It follows that (see (4.1)):

$$B \textit{ blocking set } \Longrightarrow r < |B| < v - r \tag{4.2}$$

Our aim is to improve (4.2). Let be $n = \max_{\ell \in L} |\ell \cap B|$ and s a n-secant line to B. Let be $P \in s - B$. Each of the $r-1$ lines through P, different from s, meets B in at least one point, whence $|B| \geq r - 1 + n$. Set:

$$h = |B| = r - 1 + a, \quad a \geq n = \max_{\ell \in L} |\ell \cap B| \tag{4.3}$$

If in (3.10) we set $M = 1$, $N = a$, $h = |B| = r - 1 + a$, we get:

$$\begin{cases} (r-1+a)^2 - (r-1+a)(ra+1) + ab \leq 0 \\ = 0 \iff a = n,\ B \textit{ is of type } (1,n) \end{cases} \tag{4.4}$$

Since $v = r(k-1) + 1$, $b = vr/k$, by (4.4) we get:

$$\begin{cases} ka^2 - a(3k-r) - k(r-2) \geq 0 \iff a \geq \alpha = 1 + \dfrac{k-r+\sqrt{(r-k)^2+4k(k-1)r}}{2k} \\ a = \alpha \iff a = \alpha = n,\ B \textit{ is of type } (1,n). \end{cases} \tag{4.5}$$

By (4.3), (4.5) and since the complement of a blocking set is a blocking set, we have:

$$\begin{cases} r + \dfrac{k-r+\sqrt{(r-k)^2+4k(k-1)r}}{2k} \leq |B| \leq v-r-\dfrac{k-r+\sqrt{(r-k)^2+4k(k-1)r}}{2k} \\ \text{the equality in the left hand side holds if, and only if,} \\ a = \alpha = n \text{ and } B \text{ is of type } (1,n). \text{ The equality in the right} \\ \text{hand side holds if, and only if, } B \text{ is the complement of such} \\ \text{a set.} \end{cases} \quad (4.6)$$

If S(2,k,v) is a projective plane of order q, that is if k = r = = q+1, by (4.6) we get (*Bruen*'s theorem):

I. - *In a projective plane of order q, $S(2,q+1,q^2+q+1)$, if B is a blocking set, it is:*

$$q+\sqrt{q}+1 \leq |B| \leq q^2-\sqrt{q} \quad (4.7)$$

The equality in the left hand side holds iff B is a Baer subplane, that is a $(q+\sqrt{q}+1)$-set of type $(1,1+\sqrt{q})$. The equality in the right hand side holds iff B is the complement of a Baer subplane.

If S(2,k,v) is such that r = 3k, by (4.5), (4.3) it is:

$$r = 2k \Rightarrow \begin{cases} r-1+\sqrt{r-2} \leq |B| \leq v-r+1-\sqrt{r-2} \\ \text{the equality in the left hand side holds iff } a = \\ \alpha = n - \sqrt{r-2} \text{ and } B \text{ is of type } (1,n), \text{ in the} \\ \text{right hand side, iff } B \text{ is the complement of such} \\ \text{a set.} \end{cases} \quad (4.8)$$

By (4.6), among other results, it follows that:

II. - *In S(2,3,v) blocking sets don't exist.*

Let us now consider a $S(2,c+1,c^3+1)$. The *hermitian curves* of $PG(2,q)$, q a square, provide examples of such Steiner systems, when $c = \sqrt{q}$. (A hermitian curve of $PG(2,q)$, q a square, is a curve of equation $x^{\sqrt{q}+1} + y^{\sqrt{q}+1} + z^{\sqrt{q}+1} = 0$). By (4.6) we get, if B is a blocking set of $S(2,c+1,c^3+1)$, since $r = c^2$:

$$|B| \geq c^2 + \frac{c + 1 - c^2 + \sqrt{(c^2-c-1)^2 + 4(c+1)c^3}}{2(c+1)}$$

Moreover we prove:

$$c \geq 8 \implies \frac{c + 1 - c^2 + \sqrt{(c^2-c-1)^2 + 4(c+1)c^3}}{2(c+1)} > 1 + c/2$$

It follows

$$\begin{cases} B \textit{ blocking set of } S(2,c+1,c^3+1), \ c \geq 8 \implies \\ \implies c^2 + c/2 + 1 < |B| < c^3 - c^2 - c/2 \end{cases} \tag{4.10}$$

5. EVEN TYPE SETS AND ODD TYPE SETS IN S(2,k,v)

In $S(2,k,v)$ a subset P is an *even type set* iff every line meets it in an *even* number of points. We denote by $\mathcal{P}$ the family of even type sets of $S(2,k,v)$. A subset D of $S(2,k,v)$ is an *odd type set* iff every line meets it in an *odd* number of points. We denote by $\mathcal{D}$ the family of odd type sets of $S(2,k,v)$. We set:

$$\mathcal{H} = \mathcal{P} \cup \mathcal{D} \tag{5.1}$$

It is:

$$\emptyset \in \mathcal{P}; \quad k \text{ even} \Longrightarrow S \in \mathcal{P}; \quad k \text{ odd} \Longrightarrow S \in \mathcal{D} \tag{5.2}$$

Moreover it is:

$$\begin{cases} k \text{ even} \Longrightarrow [X \in \mathcal{P} \Longrightarrow X' = S - X \in \mathcal{P}; X \in \mathcal{D} \Longrightarrow X' = S - X \in \mathcal{D}] \\ k \text{ odd} \Longrightarrow [X \in \mathcal{P} \Longleftrightarrow X' = S - X \in \mathcal{D}] \end{cases} \tag{5.3}$$

Since for any $\ell \in L$ it is:

$$X, Y \subset S \Longrightarrow |(X \cup Y - X \cap Y) \cap \ell| = |X \cap \ell| + |Y \cap \ell| - 2|X \cap Y \cap \ell|,$$

we get:

$$\begin{cases} X, Y \in \mathcal{P} \Longrightarrow (X \cup Y - X \cap Y) \in \mathcal{P}, \\ X, Y \in \mathcal{D} \Longrightarrow (X \cup Y - X \cap Y) \in \mathcal{P}, \\ X \in \mathcal{D}, Y \in \mathcal{D} \Longrightarrow (X \cup Y - X \cap Y) \in \mathcal{D} \end{cases} \tag{5.4}$$

It is known that the *power set of* S, with respect to the symmetric difference:

$$X, Y \subseteq S, \qquad X \oplus Y = X \cup Y - X \cap Y \tag{5.5}$$

and to the product times a scalar in Z_2:

$$0 \cdot X = \emptyset, \quad 1 \cdot X = X, \qquad 0,1 \in Z_2 \tag{5.5'}$$

is a *vector space* over Z_2, $(\mathbb{P}(S), \oplus, ., Z_2)$ *isomorphic to* Z_2^v $(v = |S|)$. By (5.4) we get that $(\mathcal{H}, \oplus, ., Z_2)$ is a *subspace* of $(\mathbb{P}(S), \oplus, ., Z_2)$, such that, if $\mathcal{D} \neq \emptyset$, $\mathcal{P}$ is an index two subspace.

We easily prove that:

$$\begin{cases} X \in \mathcal{P},\ X \neq S \Longrightarrow |X| \equiv 0 \bmod 2 \\ X \in \mathcal{P},\ X \neq \emptyset \Longrightarrow |X| = 1 + r \bmod 2 \end{cases} \tag{5.6}$$

Moreover:

$$\begin{cases} X \in \mathcal{D},\ X \neq S \Longrightarrow |X| \equiv r \bmod 2 \\ X \in \mathcal{D},\ X \neq \emptyset \Longrightarrow |X| \equiv 1 \bmod 2 \end{cases} \tag{5.7}$$

It follows:

I. - *If r is even the only either even type sets or odd type sets are the trivial ones, that is $\emptyset, S$. Moreover if r is odd:*

$$\begin{cases} X \in \mathcal{P},\ X \neq \emptyset, S \Longrightarrow |X| \text{ even}, \\ X \in \mathcal{D},\ X \neq \emptyset, S \Longrightarrow |X| \text{ odd}. \end{cases} \tag{5.8}$$

From now on we assume r odd:

$$r = 2\rho + 1 . \tag{5.9}$$

We easily prove:

$$D \in \mathcal{D} \Longrightarrow [\,|D| \geq r;\ |D| = r \Longleftrightarrow k \text{ odd, } D \text{ is of type } (1,k) \tag{5.10}$$
$$\Longleftrightarrow k \text{ odd, } D \text{ a prime }]$$

$$P \in \mathscr{P},\ P \neq \emptyset \Rightarrow [|P| \geq r+1;\ |P| = r+1 \Leftrightarrow P \text{ is of type } (0,2)] \qquad (5.11)$$

By (5.10), (5.8) we get:

$$k \text{ even} \Longrightarrow |D| \geq r+2, \quad D \in \mathscr{D} \qquad (5.12)$$

By (5.10), (5.11), (5.12) we have:

II.- *Every non-zero vector of* $(\mathscr{H}, \oplus, ., Z_2)$ *has weight* $w \geq r$ *if* k *is odd,* $w \geq r+1$ *if* k *is even. (We recall that the weight of* $U \in \mathscr{H}$ *is the number of non-zero components of the vector* U, *that is the number of points of* U). *It follows that* $\mathscr{H}$ *is a linear code of* Z_2^v *correcting* $e \geq$ $\geq (r-1)/2$ *errors.*

Assume now $\mathscr{D} \neq \emptyset$ and let be $D \in \mathscr{D}$. Let $|D| = 2n+1$. The equation of characters (3.1), with respect to D provides:

$$\Sigma t_{2s+1} = b,\ \Sigma(2s+1)t_{2s+1} = (2n+1)r,\ \Sigma(2s+1)2s\, t_{2s+1} = (2n+1)2n \qquad (5.13)$$

By $(5.13)_1$, $(5.13)_2$ we get:

$$2\Sigma s\, t_{2s+1} + b = 2nr + r \Longrightarrow b = r + 2\sigma$$

Therefore:

$$\Sigma s\, t_{2s+1} = nr - \sigma, \qquad b = r + 2\sigma. \qquad (5.14)$$

By $(5.13)_3$, (5.14) we get:

$$2\Sigma s^2 t_{2s+1} + nr - \sigma = 2n^2 + n$$

and then, since r is odd, it is:

$$\sigma = 2\tau \quad , \qquad b = r + 4\tau$$

Since it is vr = kb, we have:

$$vr = k(r + 4\tau) \implies r(v-k) = 4\tau k \implies \begin{cases} v \equiv k \bmod 4 \\ v \equiv k \bmod 8 \text{ if } k \text{ is even} \end{cases}$$

So we prove:

III. - *If $\mathcal{D}$ is non-empty, it is $b \equiv r \bmod 4$ and $v \equiv k \bmod 4$; it is $v \equiv k \bmod 8$ if k is even. Moreover for any $D \in \mathcal{D}$, $|D| = 2n+1$, set $b = r + 4\tau$, it is ($r = 2\rho + 1$):*

$$\Sigma s\, t_{2s+1} = nr - 2\tau, \quad \Sigma s^2\, t_{2s+1} = n^2 + \tau - n\rho. \tag{5.15}$$

By (5.15) we get ($r = 2\rho + 1$):

$$\Sigma (s^2 - s) t_{2s+1} = n(n-1) - 3n\rho - 3\tau \tag{5.16}$$

Since the left hand side in (5.16) is even, by (5.16) we get:

$$n\rho \equiv \tau \bmod 2 \qquad (b = r+4\tau,\ r(v-k) = 4\tau k) \tag{5.17}$$

If ρ is even, that is $r \equiv 1 \bmod 4$, by (5.17) we obtain that τ is even, that is $b \equiv r \bmod 8$, whence $v \equiv k \bmod 8$, $v \equiv k \bmod 16$ if k is even, that is:

IV. - *If $\mathcal{D}$ is non-empty and $r \equiv 1 \bmod 4$, it is $b \equiv r \bmod 8$, $v \equiv k \bmod 8$ and $v \equiv k \bmod 16$ if k is even.*

If ν is odd *(that is $r \equiv 3 \bmod 4$)* and τ is even, that is $b \equiv r \bmod 8$, then n is even, that is $|D| = 2n+1 \equiv 1 \bmod 4$. If τ is odd, that is $|D| = 2n+1 \equiv 3 \bmod 4$. So we prove:

V. - *If $\mathcal{D}$ is non-empty and $r \equiv 3 \bmod 4$, it is:*

$$\begin{cases} b \equiv r \bmod 8 \implies \forall D \in \mathcal{D}, \quad |D| \equiv 1 \bmod 4 \\ b \not\equiv r \bmod 8 \implies \forall D \in \mathcal{D}, \quad |D| \equiv 3 \bmod 4 \end{cases} \tag{5.18}$$

As corollaries of I, III, IV we get:

VI. - *In $S(2,q,q^m)$, q even and $m \geq 2$ (for instance in an affine plane of order q even) it is $\mathcal{D} = \emptyset$.*

VII. - *In $S(2,a+1,a^3+1)$ (abstract unital) if a is even, it is $\mathcal{D} = \{S\}$. If a is odd and $a \not\equiv 1 \bmod 8$ it is $\mathcal{D} = \emptyset$.*

VIII. - *In $S(2,n,n(cn-c+1))$ (maximal arc) it is:*

$$\begin{cases} c \text{ and } n \text{ odd} \implies \mathcal{D} = \{S\}, \\ c \text{ odd, } n \text{ even} \implies \mathcal{D} = \emptyset, \\ c \text{ and } n \text{ even, } c \not\equiv 0 \bmod 8 \implies \mathcal{D} = \emptyset. \end{cases}$$

Assume now that in $S(2,k,v)$ a non-empty even type set P exists and let k be *odd*. Then $D = S - P \in \mathcal{D}$ (see (5.3)) and by theorems III, IV, V we get:

IX. - *In $S(2,k,v)$, k odd and $\mathcal{P} \neq \{\emptyset\}$, it is $b \equiv r \bmod 4$ and $v \equiv k \bmod 4$. If $r \equiv 1 \bmod 4$ it is $b \equiv r \bmod 8$ and $v \equiv k \bmod 8$. If $r \equiv 3 \bmod 4$ it is*

$$\begin{cases} b \equiv r \bmod 8 \implies \forall P \in \mathcal{P} - \{\emptyset\}, \ |P| \equiv v-1 \bmod 4 \\ b \not\equiv r \bmod 8 \implies \forall P \in \mathcal{P} - \{\emptyset\}, \ |P| \equiv v-3 \bmod 4 \end{cases} \tag{5.19}$$

Let be $P \in \mathcal{P}$ and $|P| = 2n$. By the equations of characters referred to P, we get:

$$|P| = 2n, \quad \Sigma\, s\, t_{2s} = nr\,, \quad \Sigma\, s(2s-1)t_{2s} = n(2n-1)$$

whence (since $r = 2\rho+1$):

$$\Sigma\, s^2\, t_{2s} = n(n+\rho), \quad \Sigma\, s\, t_{2s} = n(2\rho+1)$$

so that:

$$\Sigma\, (s^2-s)t_{2s} = n(n - \rho - 1)$$

If ρ is odd, that is $r \equiv 3 \bmod 4$, since the left hand side of the previous equation is even, we get: $n \equiv 0 \bmod 2$, that is $|P| \equiv 0 \bmod 4$. So we state:

X. - In $S(2,k,v)$, $r \equiv 3 \bmod 4$, for any $P \in \mathcal{P}$, it is $|P| \equiv 0 \bmod 4$.

If $r \equiv 3 \bmod 4$ and k is odd, by prop. X and (5.3) we get:

$$D \in \mathcal{D} \iff S - D \in \mathcal{P} \implies |S-D| \equiv 0 \bmod 4 \implies |D| \equiv v \bmod 4.$$

By this and by prop. V we get:

XI. - In $S(2,k,v)$, $r \equiv 3 \bmod 4$ and k odd, if $\mathcal{D} \neq \emptyset$, for any $D \in \mathcal{D}$ it is $|D| \equiv v \bmod 4$, whence:

$$\begin{cases} b \equiv r \bmod 8 \implies v \equiv 1 \bmod 4, \\ b \not\equiv r \bmod 8 \implies v \equiv 3 \bmod 4. \end{cases} \qquad (5.20)$$

6. LINEAR CODES RELATED TO A S(2,k,v), r ≡ 3 mod 4

Let S(2,k,v) = (S,L) be a Steiner system with $r \equiv 3 \bmod 4$. As we previously considered in sect. 5 we associate to it the subspace $\mathscr{H} = \mathscr{P} \cup \mathscr{D}$ of the vector space $\mathbb{P}(S) = (\mathbb{P}(S), \oplus, \cdot, Z_2)$, that is a *linear (v,w,d)-code* (where $v = \dim \mathbb{P}(S) = |S|$, w = weight of $\mathscr{H} \geq r$, $d = \dim \mathscr{H}$). Our aim is now to calculate d.

In $\mathbb{P}(S)$ we define a *scalar product* in the classical way:

$$X,Y \in \mathbb{P}(S), \quad X = (x_i),\ Y = (y_i),\ X.Y = \Sigma x_i y_i \quad (\in Z_2)$$

We easily prove:

$$X,Y \in \mathbb{P}(S), \quad X.Y = |X \cap Y|_2 = \begin{cases} = 0 & \text{if } |X \cap Y| \text{ is even,} \\ = 1 & \text{if } |X \cap Y| \text{ is odd.} \end{cases}$$

So in $\mathbb{P}(S)$ the following orthogonality is defined:

$$X \perp Y \iff |X \cap Y|_2 = 0, \qquad X,Y \in \mathbb{P}(S) \tag{6.1}$$

If T is a subspace of $\mathbb{P}(S)$ we set:

$$T^{\perp} = \{X \in \mathbb{P}(S) : X \perp Y,\ \forall\, Y \in T\}. \tag{6.2}$$

We easily prove:

$$\dim T + \dim T^{\perp} = \dim \mathbb{P}(S) = v. \tag{6.3}$$

By proposition X we get:

$$\begin{cases} X\in\mathcal{P} \Rightarrow \forall Y\in\mathcal{P},\ |X \oplus Y| = |X\cup Y - X\cap Y| = |X|+|Y|-2|X\cap Y| \equiv 0 \bmod 4 \\ |X| \equiv 0 \bmod 4,\ |Y| \equiv 0 \bmod 4 \Rightarrow |X\cap Y| \equiv 0 \bmod 2 \Rightarrow X\perp Y,\ \forall Y\in P \end{cases} \tag{6.4}$$

Let a = 1 if b ≡ r mod 8, a = 3 if b ≢ r mod 8. By prop. V we get:

$$\begin{cases} X\in\mathcal{P} \Rightarrow \forall Y\in\mathcal{D},\ |X \oplus Y| = |X|+|Y|-2|X\cap Y| \equiv a \bmod 4, \\ |X| \equiv 0 \bmod 4,\ |Y| \equiv a \bmod 4 \Rightarrow |X\cap Y| \equiv 0 \bmod 2 \Rightarrow X\perp Y, \forall Y\in\mathcal{D} \end{cases} \tag{6.5}$$

By (6.4), (6.5) we get:

$$X\in\mathcal{P} \Rightarrow \forall Y\in\mathcal{H},\ X\perp Y \iff X\in\mathcal{H}^{\perp} \tag{6.6}$$

that is:

$$\mathcal{P}\subseteq\mathcal{H}^{\perp} \Rightarrow \dim\mathcal{P} \le \dim\mathcal{H}^{\perp} \tag{6.7}$$

If $\mathcal{D}$ = Ø, that is $\mathcal{P}=\mathcal{H}$, by (6.7) and (6.3) we get: $2\dim\mathcal{P} \le \dim\mathcal{P} + \dim\mathcal{P}^{\perp} = v$, that is

$$\mathcal{D} = \emptyset \Rightarrow d \le v/2 \tag{6.8}$$

If $\mathcal{D}$ ≠ Ø, whence dim $\mathcal{P}$ = d-1, by (6.7) and (6.3) we get: 2d-1 = = dim $\mathcal{H}$ + dim $\mathcal{P}$ ≤ dim $\mathcal{H}$ + dim $\mathcal{H}^{\perp}$ = v, that is:

$$\mathcal{D} \neq \emptyset \Rightarrow d \le (v+1)/2 \tag{6.9}$$

Assume now $L\subset\mathcal{H}$, whence $L\subset\mathcal{D}$ and k is odd, moreover S(2,k,v) is a projective plane of order q = k-1, that is a $S(2,q+1,q^2+q+1)$, with

$q \equiv 2 \bmod 4$ (since $r = q+1 \equiv 3 \bmod 4$). We have:

$$X \in \mathcal{H}^{\perp} \Rightarrow \forall y \in \mathcal{H},\ X \perp y \iff |X \cap y| \equiv 0 \bmod 2,\ \forall y \in \mathcal{H}$$
$$\Rightarrow \forall \ell \in L,\ |X \cap \ell| \text{ is even} \iff X \in \mathcal{P},$$

that is:

$$\mathcal{H}^{\perp} \subseteq \mathcal{P} \Rightarrow \dim \mathcal{H}^{\perp} \leq \dim \mathcal{P} \tag{6.10}$$

By (6.10) and (6.3) we get: $v = \dim \mathcal{H} + \dim \mathcal{H}^{\perp} \leq \dim \mathcal{H} + \dim \mathcal{P} = 2d-1 \Rightarrow$
$\Rightarrow d \geq (v+1)/2$, that is:

$$L \subseteq \mathcal{D} \Rightarrow d \geq (v+1)/2 \tag{6.11}$$

By (6.9) and (6.11) we obtain:

I. - *In a projective plane $S(2,q+1,q^2+q+1)$ with $q \equiv 2 \bmod 4$, it is $d = (q^2+q+2)/2$ and $\mathcal{H}$ is a linear $(v=q^2+q+1,\ w=q+1,\ d=(q^2+q+2)/2)$-code.*

At least we prove:

II. - *In $PG(m,2) = S(2,3,\theta_m)$, $\mathcal{H}$ is a linear $(v=\theta_m,\ w=\theta_{m-1},\ m+2)$-code.*

Proof. In $PG(m,2)$ every odd type set is either $PG(m,2)$ or a hyperplane (since it is of class $[1,3]$). Then every even type set is either $\emptyset$ or the complement of a hyperplane. It follows $w = \theta_{m-1}$ and $|\mathcal{H}| = 2\theta_m + 2 = 2^{m+2}$ and then $d = m+2$.

For example by II we have:

$m = 2 \Rightarrow (7,3,4)$-code

$m = 3 \Rightarrow (15,7,5)$-*code*

$m = 4 \Rightarrow (31,15,6)$-*code*

$m = 5 \Rightarrow (63,31,7)$-*code*

Theorems I and II are well known but here they follow as particular cases of the general geometric theory previously explained.

REFERENCES

[1] A. Beutelspacher, Blocking sets and partial spreads in finite projective spaces, Math. Z., 145 (1975), 211-229.

[2] A.A. Bruen, Baer subplane and blocking set, Bull. Amer. Math. Soc., 76 (1970), 342-344.

[3] A.A. Bruen, Blocking sets in projective planes, Siam. J. Appl. Math. 3 (1971), 380-392.

[4] A.A. Bruen and J.A. Thas, Blocking sets, Geometriae Dedicata, 6 (1977), 193-203.

[5] P.J. Cameron and J.H. van Lint, Graph theory, coding theory and block designs, L M S Lecture Note Series 19, 1975, Cambridge Univ. Press.

[6] P.V. Ceccherini, G. Tallini, Codes, caps and linear spaces, Finite geometries and designs, Proc. of the second Isle of Thorns Conference 1980, London Math. Soc. Lecture Note Series 49, Cambridge Univ. Press 1981.

[7] P.V. Ceccherini, G. Tallini, Caps related to incidence structures and to linear codes, Annals of Discrete Math., North Holland Publ. Co. 14 (1982) 175-182.

[8] P. Dembowsky, Finite geometries, Ergebnisse der Math. Springer, Berlin, 1968.

[9] J. Doyen, A. Rosa, An updated bibliography and survey of Steiner systems, Annals of Discrete Math., 7 (1980), 317-349.

[10] J.W.P. Hirschfeld, Projective Geometries over Finite Fields, Clarendon Press, Oxford (1979).

[11] F.J. Mac Williams, N.J.A. Sloane, The theory of Error-correcting Codes, North Holland Publ. Co. Amsterdam, New York, Oxford (1977).

[12] F. Mazzocca, G. Tallini, On the non existence of blocking sets in PG(n,q) and AG(n,q) for all large enough n, Simon Stevin 1 (1985), 43-50.

[13] B. Segre, Ovals in a finite projective plane, Canad. J. Math., 7 (1955) 415-416.

[14] B. Segre, Curve razionali normali e k-archi negli spazi finiti, Ann. Mat. pura appl. (4) 39 (1955) 357-379.

[15] B. Segre, Le geometrie di Galois, Ann. Mat. (4) 48 (1959) 1-97.

[16] B. Segre, Introduction to Galois geometries, Mem. Acc. Naz. Lincei, (8) 8 (1967) 133-236.

[17] G. Tallini, Le geometrie di Galois e le loro applicazioni alla statistica e alla teoria dell'informazione, Rend. Mat. 19 (1960) 379-400.

[18] G. Tallini, On caps of kind s in a Galois r-dimensional space, Acta Arithm. VII (1961) 19-28.

[19] G. Tallini, Un'applicazione della geometria di Galois a questioni di statistica, Rend. Acc. Naz. Lincei, (8) 35 (1963) 479-485.

[20] G. Tallini, Problemi e risultati sulle geometrie di Galois, Relaz. n. 30, Ist. Mat. Univ. Napoli, 1973.

[21] G. Tallini, Graphic characterization of algebraic varieties in a Galois space, Atti Conv. Teorie Combinatorie (Rome, sept. 1973), Acc. Naz. Lincei, 1976, 1-7.

[22] G. Tallini, Codici e geometrie combinatorie, Quaderno seminario Geom. Comb. n. 23, Ist. Mat. G. Castelnuovo Univ. Roma, marzo 1980.

[23] G. Tallini, k-insiemi e blocking sets in PG(r,q) e in AG(r,q), Quaderno n. 1 Sem. Geom. Comb. Ist. Mat. Applicata, Fac. Ingegneria Univ. L'Aquila (1982), 1-36.

[24] G. Tallini, Blocking sets nei sistemi di Steiner e d-blocking sets in PG(r,q) ed AG(r,q), Quaderno n. 3 Sem. Geom. Comb. Ist. Mat. Applicata, Fac. Ingegneria Univ. L'Aquila (1983), 1-32.

[25] G. Tallini, Teoria dei k-insiemi in uno spazio di Galois. Teoria dei codici correttori, Quaderno n. 68 Sem. Geom. Comb. maggio 1985, Dip. Mat. Univ. Roma "La Sapienza" (1985), 1-141.

[26] G. Tallini, Spazi parziali di rette e codici correttori, Rivista di Mat. pura e appl. Univ. di Udine (1987), 43-69.

[27] G. Tallini, Linear codes associated with geometric structures, Results in Math. Birkhäuser Verlag, Basel (1987), 411-422.

[28] G. Tallini, On blocking sets in finite projective and affine spaces, Annals of Discrete Math. 37 (1988), 433-450.

[29] M. Tallini Scafati, Calotte di tipo (m,n) in uno spazio di Galois $S_{r,q}$, Rend. Acc. Naz. Lincei (8) 53 (1972), 71-81.

[30] M. Tallini Scafati, Sui k-insiemi di uno spazio di Galois $S_{r,q}$ a due soli caratteri nella dimensione d, Rend. Acc. Naz. Lincei (8) 60 (1976), 782-787.

CIRCLE GEOMETRY AND ITS APPLICATION TO CODE THEORY

H. Karzel
Technical University, München, F.R.G.

ABSTRACT

There are close relations between the different types of finite circle geometries, like Möbius-, Laguerre- or Minkowski planes, and binary codes via incidence matrices as well as arbitrary q-nary codes via chain structures. In certain cases these q-nary codes are (MDS)- codes and Laguerre codes. Many geometric problems can be translated into the language of codes and vice versa. The objective of this lecture is to discuss these various aspects. The question how all these structures (or which of these) can be used for practical purposes in coding or in cryptography remains for further research.

INTRODUCTION

Coding theory can be considered as a part of finite geometry and many structures which are studied in finite geometry found and will find their applications in the theory of codes and cryptography [1]). It is the task of my lecture to consider finite circle geometries in this frame. Roughly speaking, circle geometries are characterized by the property that in general three distinct points can be

[1]) I like to mention that it is to a great deal to the credit of W. HEISE, who realized and made known the close relations between these two subjects in several papers and two excellent books, one [1] published 1976 under the coauthorship with H.-R. HALDER, the other [2] published 1983 with P. QUATTROCCHI.

joined by exactly one block, and in this case the blocks are called circles. Since the derivation of a circle geometry in a point p (cf. §1 and §3) is an incidence structure where two points determine one block, we have also to study incidence spaces and special chain-structures (cf. §3).

In the first part of this paper we review the connections between codes and finite incidence structures. After a compilation of the basic definitions and facts for incidence structures, in particular for incidence spaces (§1) and finite incidence structures (§2) we define in §3 the notions "web", "chain structure" and "cicular structure". The webs were introduced by W. BLASCHKE 1928 for purposes in Differential Geometry (cf.[3] §15). To the circular structures belongs the class of circle planes which comprises the Möbius-, the Laguerre- and the Minkowski planes [2]). The chain structures are a generalization of the Laguerre- and Minkowski planes, and incidence structures derived from nearfields, or more general, from neardomains (cf. [5], [6], [3] §§ 17, 19). There is a one to one correspondence between binary codes and the blocks of a finite incidence structure via incidence matrices (§4) and there are also close relations between webs, chain structures and q-nary codes. In order to characterize geometrically q-nary codes we need the notion "chain structure"; for each q-nary code can be identified with the chains $\mathfrak{K}$ of a 1-chain structure $(P, \mathfrak{G}, \mathfrak{K})$.

For the class of incidence spaces we introduce also the notions "m-line", "parallelism" and "double space" (§1). According to a recent observation of B. REINMIEDL certain circle geometries can be derived from double spaces so that these structures could be of some interest in the future, also in coding theory.

In the sections 5, 6 and 7 we have a closer look at incidence spaces where each line has the same cardinality. We make assumptions about the existence of m-lines. This leads us to the notion of a uniform book, the possibility of classifications of incidence spaces and problems, concerning the existence of incidence spaces with certain types of uniform books. Some of these problems can be solved as soon as it is possible to construct certain (MDS)-codes (§8). However not all problems can be tackled with (MDS)-codes, we need the more general concept of an m-incidence code (§9). (MDS)-codes can be described by regular permutation sets (§10).

The last part (§§ 11, 12, 13) is devoted to various circular structures and their

2) The reader finds a comprehensive presentation of circle planes in the standard work of W. BENZ [4].

connections with codes. The circular structures which were studied already at the time of Euclid are the circles in the Euclidean plane (cf. [3] p. 95). In the last century A.F. Möbius explored the structure of their automorphism group and in 1935 B.L. VAN DER WAERDEN and L.J. SMID gave an axiomatization for these circle planes. If one considers only the incidence axioms one obtains the notion "Möbius plane", a concept, which includes also finite structures.

Further generalizations are Möbius spaces and Möbius structures (§11). But the only finite Möbius spaces are the finite Möbius planes. To the finite Möbius structures we can associate binary codes via incidence matrices.

The abstract notion of a Laguerre plane (cf. [3] §19) has its origin in a paper of E. LAGUERRE from 1853 and the paper of VAN DER WAERDEN, SMID from 1935. The Laguerre planes are a subclass of the Laguerre structures (§12) and Laguerre structures can be described as special 1-chain structures. Since any finite 1-chain structure can be identified with a q-nary code (cf. §4), we are led to the notion "Laguerre code". The derivation of a Laguerre code is an (MDS)-code.

If one considers "circles" in a 2-dimensional space-time-world, one obtains the abstract notion of a Minkowski plane (cf. [3] §17), which includes also finite models. In a natural way each Minkowski plane can be turned into a 1-chain structure and so into a q-nary code $\mathfrak{C}$, called Minkowski code (§13). The derivation of a Minkowski plane in a point is a 2-structure, and the derivation of the corresponding q-nary Minkowski code is an (MDS)-code of length q; a q-nary Minkowski code has the length $q+1$. A large class of Minkowski codes can be derived from finite KT-nearfields. These codes have then a quite large automorphism group. Of course also the finite Laguerre structures and finite Minkowski planes determine binary codes via incidence matrices.

1. Definitions and Basic Facts about Incidence Structures in Particular Incidence Spaces

Let $\mathbb{N} := \{1,2,3,\dots\}$ be the set of all natural numbers and $\mathbb{N}_0 := \mathbb{N} \cup \{0\}$. For any set M we denote by $\mathfrak{P}(M)$ the power set of M (i.e. the set of all subsets of M including the empty set $\emptyset$ and the whole set M) and by $|M|$ the cardinal number of M. For $i \in \mathbb{N}_0$ let $\binom{M}{i} := \{X \in \mathfrak{P}(M) \mid |X| = i\}$ be the set of all i-sets of M.

Now let us consider a non empty set $\mathbf{P}$ whose elements will be called **points** and a subset $\mathfrak{B}$ of $\mathfrak{P}(P)$ whose elements will be called **blocks.** We say that a point $p \in P$ is **incident** with a block $B \in \mathfrak{B}$, if $p \in B$. The pair $(P,\mathfrak{B})$ is called an **incidence structure** and the elements of the set $I := \{(p,B) \in P \times \mathfrak{B} \mid p \in B\}$ are called **flags.**

For $D \subset P$ let $[D] := \{B \in \mathfrak{B} \mid D \subset B\}$ be the set of all blocks containing D, $P^{(D)} := P \setminus D$ and $\mathfrak{B}^{(D)} := \{B \setminus D \mid B \in [D]\}$. Then the incidence structure $\left(P^{(D)},\mathfrak{B}^{(D)}\right)$ is called the **derivation** of $(P,\mathfrak{B})$ in the set D.

For an incidence structure $(P,\mathfrak{B})$ we introduce the following numerical functions and parameters so far as they are finite. Let $i \in \mathbb{N}_0$; then

$k^{(i)}: \binom{\mathfrak{B}}{i} \to \mathbb{N}_0; \quad \{B_1,\dots,B_i\} \to |B_1 \cap B_2 \cap \dots \cap B_i|$

$r^{(i)}: \binom{P}{i} \to \mathbb{N}_0; \quad \{x_1,\dots,x_i\} \to |[\{x_1,x_2,\dots,x_i\}]|.$

For $i = 0$, $v := k^{(0)} = |P|$ is the number of points and $b := r^{(0)} = |\mathfrak{B}|$ the number of blocks of our incidence structure $(P,\mathfrak{B})$.

An important class of incidence structures are the **incidence spaces** $(P,\mathfrak{B})$ characterized by $r^{(2)}$ is constant $= 1$ and $k^{(1)} \geq 2$. (Many facts of this paragraph can be found in [7], [8], [9].) In this case the blocks $L \in \mathfrak{L}$ are called **lines.** For $a,b \in P$ with $a \neq b$ we denote by $\overline{a,b}$ the line of $\mathfrak{L}$ uniquely determined by $a,b \in \overline{a,b}$. A subset $S \subset P$ is called a **subspace** if $x,y \in S$, $x \neq y$ implies $\overline{x,y} \subset S$. Let $\mathfrak{T}$ be the **set of all subspaces** of $(P,\mathfrak{L})$. Then $\mathfrak{T}$ is $\cap$-closed and so we can define the closure operation $\overline{}: \mathfrak{P}(P) \to \mathfrak{T}$; $X \to \overline{X} := \cap \{T \in \mathfrak{T} \mid X \subset T\}$ which is called **incidence closure** or **incidence hull.** For $T \in \mathfrak{T}$ the cardinal number $\dim T := \inf \{|X| \mid \overline{X} = T\} - 1$ is called the **dimension** of T, and if $\dim T \geq 2$ we define $\mathfrak{L}(T) := \{L \in \mathfrak{L} \mid L \subset T\}$ and for $t \in T$, $\mathfrak{L}(t,T) := [t] \cap \mathfrak{L}(T) = \{L \in \mathfrak{L} \mid t \in L \subset T\}$; then $(T,\mathfrak{L}(T))$ is an incidence space. Let $\mathfrak{E} := \{E \in \mathfrak{T} \mid \dim E = 2\}$ be the set of all **planes.** A subset $S \subset P$ is called **collinear** if $\dim \overline{S} \leq 1$. An incidence space $(P,\mathfrak{L})$ is an **exchange space** if the axiom

A. $S \subset P$, $x,y \in P$ with $x \in \overline{S \cup \{y\}} \setminus \overline{S}$ implies $y \in \overline{S \cup \{x\}}$.

is valid.

Clearly every subspace $T \in \mathfrak{T}$ with $\dim T \leq 1$ is an exchange space. A plane $E \in \mathfrak{E}$ is called a **minimal plane** if one of the equivalent conditions is valid:

(a) $F \in \mathfrak{E}$ and $F \subset E$ implies $F = E$.

(b) E is spanned by any three non-collinear points contained in E.

(c) $(E, \mathfrak{L}(E))$ is an exchange space.

Let $\mathfrak{E}_m$ be the set of all minimal planes.

Two lines $A, B \in \mathfrak{L}$ are called **natural parallel,** denoted by $A \parallel B$, if either $A = B$ or $A \cap B = \emptyset$, $B \subset \overline{A \cup \{b\}}$ for all $b \in B$ and $A \subset \overline{B \cup \{a\}}$ for all $a \in A$.
For $L \in \mathfrak{L}$, $a \in P \setminus L$ let $(a \parallel L) := \{X \in \mathfrak{L}(a, \{a\} \cup L) \mid X \cap L = \emptyset\}$, and if $m \in \mathbb{N}_0$ then a line $L \in \mathfrak{L}$ is called an **m-line,** if for all $x \in P \setminus L$, $|(x \parallel L)| = m$ and, in the case $m = 0$ further $|L| \geq 3$. Let $\mathfrak{L}_m$ be the set of all m-lines. The lines of $\mathfrak{L}_0$ resp. $\mathfrak{L}_1$ are also named **projective** resp. **affine lines.**

An incidence space $(P, \mathfrak{L})$ is called

a **projective space** if $\mathfrak{L} = \mathfrak{L}_0$; here it follows that the function $k^{(1)}(L) = k$ is constant; $n := k - 1$ is called the **order** of the projective space,

a **pseudo-affine space** if $\mathfrak{L} = \mathfrak{L}_1$; here it follows also that $k^{(1)}(L) = k$ is constant, (k is called the **order** of the pseudoaffine space), and that if $k \geq 4$, the natural parallelism $\parallel$ is transitive (cf. [10]),

an **affine space** if $\mathfrak{L} = \mathfrak{L}_1$ and if $\parallel$ is transitive,

an **m-porous space** if $\mathfrak{L} = \mathfrak{L}_0 \cup \mathfrak{L}_1 \cup \ldots \cup \mathfrak{L}_m$.

(1.1) Let $(P, \mathfrak{L})$ be an incidence space with $\mathfrak{L}_1 \neq \emptyset$, and let $A \in \mathfrak{L}_1$ and $x, y \in P \setminus A$; then:

a) $X := (x \parallel A)$ is a single line and $\exists\, a \in A: |\overline{x,a}| \geq 3$.

b) If $a \in A$, $b \in \overline{a,x} \setminus \{a, x\}$ and $B := (b \parallel A)$ then $B \cap X = \emptyset$.

c) If $a \in A$ and $b \in a,x \setminus \{a, x\}$ then $\forall y \in X: \overline{b,y} \cap A \neq \emptyset$, and so $[X \overset{b}{\rightarrow} A]: X \rightarrow A;\ y \rightarrow \overline{b,y} \cap A$ is an injection.

d) $|X| \leq |A|$.

e) $\forall y \in X: (y \parallel A) = X$.

f) If $X \in \mathfrak{L}_1$ then $\forall a \in A: (a \parallel X) = A$, $X \parallel A$ and $|X| = |A|$.

Proof. a) Clearly $X \not\subset M := \bigcup \{\overline{x,a} \mid a \in A\} \subset \overline{\{x\} \cup A} =: E$ but $X = (x \parallel A) \subset E$. If $\overline{x,a} = \{x, a\}$ for all $a \in A$ then $M \in \mathfrak{T}$, hence $M = \overline{\{x\} \cup A} = E$. Therefore there is an $a \in A$ with $|\overline{x,a}| \geq 3$.

b) Let $b \in \overline{x,a} \setminus \{x, a\}$ and $B := (b \parallel A)$. Then $B \subset \overline{\{b\} \cup A} = E$. Suppose $B \cap X \neq \emptyset$

and let $c \in B \cap X$; then $c \neq b$. By **a**) there is an $a' \in A$ such that $|\overline{c,a'}| \geq 3$; let $b' \in \overline{c,a'} \setminus \{c,a'\}$. Since $c \neq b$, we have $b' \neq b$, $b' \notin B$ and $\overline{b,b'} \subset E$ implying $\overline{b,b'} \cap A \neq \emptyset$. This gives us $E = \overline{b' \cup A} = \overline{c \cup A}$, hence $X, B \in (c \parallel A)$, hence $X = B$ because $A \in \mathfrak{L}_1$ and so the contradiction $b = x$.

c) Let $B := (b \parallel A)$. By **b**) $X \cap B = \emptyset$ and since $E := \overline{\{x\} \cup A} = \overline{\{b\} \cup A}$ we have $\overline{b,y} \subset E$, $\overline{b,y} \neq X, B$ for each $y \in X$ and so $\overline{b,y} \cap A \neq \emptyset$ since $A \in \mathfrak{L}_1$.

d) is a consequence of **c**).

e) Let $a \in A$ such that $|\overline{x,a}| \geq 3$, let $b \in \overline{x,a} \setminus \{x,a\}$ and $B := (b \parallel A)$. By **c**) $\overline{b,y} \cap A \neq \emptyset$, hence $E := \overline{\{x\} \cup A} = \overline{\{b\} \cup A} = \overline{\{y\} \cup A}$ and so $y \in X$, $(y \parallel A) \subset \overline{\{y\} \cup A}$. Therefore $(y \parallel A) = X$.

f) Let again $E := \overline{\{x\} \cup A}$ and $a' \in A$ such that $|\overline{x,a'}| \geq 3$ (cf. **a**)). Then by **c**) $E = \overline{\{a'\} \cup X}$, hence $A = (a' \parallel X)$, and since $X \in \mathfrak{L}_1$ this is true for all $a \in A$ by **e**), i.e. $(a \parallel X) = A$ and $\overline{\{a\} \cup X} = E$ for all $a \in A$, consequently $X \parallel A$ and finally $|X| = |A|$ by **d**).

By a **parallelism** $\parallel'$ we understand an equivalence relation on the set $\mathfrak{L}$ of lines such that **Euclid's parallelity-axiom** is valid in the following form:

(E.P) $\forall a \in P,\ \forall L \in \mathfrak{L}\ \exists_1 M \in [a]$ such that $M \parallel' L$; we denote then $(a \parallel' L) := M$.

The relation "natural parallel" is a parallelism if and only if $(P, \mathfrak{L})$ is an affine space. But there are incidence spaces $(P, \mathfrak{L})$ different from affine spaces which do admit a parallelism $\parallel'$ and then of course $\parallel' \neq \parallel$. There are even affine spaces with parallelisms which are different from the natural parallelisms.

A special class of incidence spaces that admit parallelisms are the **double spaces** $(P, \mathfrak{L}, \parallel_l, \parallel_r)$. Here $\parallel_l$ and $\parallel_r$ are parallelisms, not necessarily distinct, such that:

(D) For $A, B \in \mathfrak{L}$, $a \in A$, $b \in B$: $A \cap B \neq \emptyset \iff (a \parallel_l B) \cap (b \parallel_r A) \neq \emptyset$.

The class of all 2-porous double spaces with $k^{(1)} \geq 9$ is completely determined (cf. [11]); these double spaces can be derived from certain algebraic structures.

For a line $A \in \mathfrak{L}$ we consider the two sets of planes $\mathfrak{E}(A) := \{E \in \mathfrak{E} \mid A \subset E\}$ and $\mathfrak{E}_s(A) := \{\overline{A \cup \{x\}} \mid x \in P \setminus A\}$. Clearly $\mathfrak{E}_s(A) \subset \mathfrak{E}(A)$ and $\cup\, \mathfrak{E}_s(A) = P$. $\mathfrak{E}(A)$ resp. $\mathfrak{E}_s(A)$ is called a **book** resp. a **small book** with the **spine** A if for all $E, F \in \mathfrak{E}(A)$ resp. $E, F \in \mathfrak{E}_s(A)$ with $E \neq F$, $E \cap F = A$. If $\mathfrak{E}(A)$ is a book, then $\mathfrak{E}_s(A) = \mathfrak{E}(A)$. Let $\mathfrak{L}_s := \{A \in \mathfrak{L} \mid \mathfrak{E}_s(A)$ is a small book$\}$ be the set of all lines which are spines of small books. If $\mathfrak{E}_s(A) \subset \mathfrak{E}_m$ then $A \in \mathfrak{L}_s$. Let $A \in \mathfrak{L}$, $p,q \in P\setminus A$ with $p \neq q$. If $\overline{p,q} \cap A \neq \emptyset$ then $\overline{A \cup \{p\}} = \overline{A \cup \{q\}}$. Therefore if $A \in \mathfrak{L}_0$ then $q \in \overline{A \cup \{p\}}$ implies $\overline{A \cup \{p\}} = \overline{A \cup \{q\}}$, hence $\mathfrak{L}_0 \subset \mathfrak{L}_s$. If $A \in \mathfrak{L}_1$, $\overline{p,q} \cap A = \emptyset$ and $q \in \overline{A \cup \{p\}}$ then $q \in (p \parallel A)$ and so $p \in (p \parallel A) = (q \parallel A) \subset \overline{A \cup \{q\}}$ by (1.1.e). This shows $\mathfrak{L}_1 \subset \mathfrak{L}_s$. Hence

(1.2) $\mathfrak{L}_0 \cup \mathfrak{L}_1 \subset \mathfrak{L}_s$.

To each point $p \in P$ we can associate the following **bundle structure** $([p], \mathfrak{E}_s(p))$, where $\mathfrak{E}_s(p) := \{\overline{\{p\} \cup L} \mid L \in \mathfrak{L}\setminus[p]\}$ is the set of all planes which are generated by p and a line L not incident with p. A new point $X \in [p]$ (i.e. a line X with $p \in X$) is incident with a new block $E \in \mathfrak{E}_s(p)$ if and only if $X \subset E$.

(1.3) For a point $p \in P$ the following statements are equivalent:

I a) $[p] \subset \mathfrak{L}_s$.

Ib) The bundle structure $([p], \mathfrak{E}_s(p))$ is an incidence space (and then called **bundle space**).

Proof. Let $X \in [p]$ and $E,F \in \mathfrak{E}_s(X)$ with $E \subset F$; then $E,F \in \mathfrak{E}_s(p)$ and there is an $Y \in [p]\setminus\{X\}$ with $Y \subset E \subset F$. If $([p], \mathfrak{E}_s(p))$ is an incidence space then $E = F$ and so $X \in \mathfrak{L}_s$. Hence $[p] \subset \mathfrak{L}_s$.

Now let $[p] \subset \mathfrak{L}_s$ and $X,Y \in [p]$ with $X \neq Y$; then $E := \overline{X \cup Y} \in \mathfrak{E}_s(p)$. Let $F \in \mathfrak{E}_s(p)$ with $X,Y \subset F$; then $E \subset F$ and there are $A,B \in [p]$ with $F = \overline{A \cup B}$. Suppose $X \neq A$; then $F, \overline{X \cup A} \in \mathfrak{E}_s(A)$ and $\overline{X \cup A} \subset F$, hence $F = \overline{X \cup A}$ since $A \in \mathfrak{L}_s$. Consequently, $E,F \in \mathfrak{E}_s(X)$ and so $E = F$ since $X \in \mathfrak{L}_s$. Therefore $([p], \mathfrak{E}_s(p))$ is an incidence space.

2. Finite Incidence Structures

Let $(P,\mathfrak{B})$ be a finite incidence structure, i.e. $v = k^{(0)} = |P| \in \mathbb{N}_0$ and then also $b = r^{(0)} = |\mathfrak{B}| \in \mathbb{N}_0$. By counting the flags of I in two different ways (each point $p \in P$ is incident with $r^{(1)}(p)$ blocks and each block $B \in \mathfrak{B}$ is incident with $k^{(1)}(B)$ points) we obtain the formula:

(2.1) $\sum_{p \in P} r^{(1)}(p) = |I| = \sum_{B \in \mathfrak{B}} k^{(1)}(B)$.

If $k\ (B) = k$ is constant for all blocks $B \in \mathfrak{B}$, i.e. all blocks have the same size, then by (2.1) $|I| = b \cdot k$, and if $r^{(1)}(p) = r$ is constant for all $p \in P$ then $|I| = v \cdot r$.

(2.2) For a **tactical configuration** $(P,\mathfrak{B})$, i.e. both functions $k^{(1)} = k$ and $r^{(1)} = r$ are constant, we have $v \cdot r = b \cdot k$.

For a subset $D \subset P$ the derivation $(P', D') = (P^{(D)}, \mathfrak{B}^{(D)})$ has the parameters $v' = v - |D|$ and $b' = r^{(|D|)}(D)$. Let $i := |D| + 1$ and assume that the functions $k^{(1)} = k$ and $r^{(i)}$ are constant; then $k^{(1)} = k \setminus |D| = k + 1 - i$ and $r' = r^{(i)}$ are constant, i.e. the derivation $(P^{(D)}, \mathfrak{B}^{(D)})$ is a tactical configuration, and by **(2.2)** we have:

$$(v + 1 - i) \cdot r^{(i)} = b' \cdot (k + 1 - i) = r^{(i-1)}(D) \cdot (k + 1 - i).$$

This implies that also $r^{(i-1)}$ is constant. Therefore we have the theorem:

(2.3) Let $(P,\mathfrak{B})$ be a finite incidence structure.

a) If there is an $i \in \mathbb{N}$ such that the two functions $k^{(1)} = k$ and $r^{(i)}$ are constant then also the functions $r^{(j)}$ for $0 \leq j < i$ are constant, hence $(P,\mathfrak{B})$ is a tactical configuration and we have $r^{(i-1)} = \frac{v+1-i}{k+1-i} r^{(i)}$, hence $b = r^{(o)} = \left(\prod_{j=1}^{i} \frac{v+1-j}{k+1-j} \right) r^{(i)}$.

b) If for an $i \in \mathbb{N}$ the functions $k^{(i)}$ and $r^{(1)} = r$ are constant then also the functions $k^{(j)}$ for $0 \leq j < i$ are constant, hence $(P,\mathfrak{B})$ is a tactical configuration and $k^{(i-1)} = \frac{b+1-i}{r+1-i} k^{(i)}$, hence $v = k^{(o)} = \left(\prod_{j=1}^{i} \frac{b+1-j}{r+1-j} \right) k^{(i)}$.

Remarks. We say that an incidence structure $(P,\mathfrak{B})$ (not necessarily finite) is of type $\mathbf{T(t,\lambda,k)}$, $t,\lambda,k \in \mathbb{N}$ if $k^{(1)} = k$ and $r^{(t)} = \lambda$ are constant, i.e. each block is incident with exactly k points and each set of t distinct points is contained in exactly λ blocks. If $(P,\mathfrak{B})$ is of type $T(t,\lambda,k)$ with $t,\lambda,k > 1$ then for each point $a \in P$ the derivation $(P^a, \mathfrak{B}^a)$ is of type $T(t-1, \lambda, k-1)$. We define $\mathbf{S(t,k)} := T(t,1,k)$ and for $v \in \mathbb{N}$, $T(t,\lambda,k,v) := \{(P,\mathfrak{B}) \in T(t,\lambda,k) \,|\, |P| = v\}$, $S(t,k,v) := T(t,1,k,v)$. By (2.3) the incidence structures of $T(t,\lambda,k,v)$ are tactical configurations. The incidence structures of type $S(t,k,v)$ are called **Steiner systems.** $S(2,k)$ denotes the set of all incidence spaces $(P,\mathfrak{L})$ where each line $L \in \mathfrak{L}$ is incident with exactly k points.
By (2.3) we obtain:

(2.4) If $(P,\mathfrak{B})$ is of type $S(t,k,v)$ then for each $i \in \{0,1,\ldots,t-1\}$ the function

$r^{(i)} = \prod_{j=i}^{t-1} \frac{v-j}{k-j}$ is a constant integer. For $t = 2$ the numbers $r := r^{(1)} = \frac{v-1}{k-1}$ and $b = r^{(0)} = \frac{v}{k} \cdot \frac{r-1}{k-1}$ are integers, hence $v = 1 + r(k-1)$ and $b = \frac{(1+r(k-1))r}{k}$, i.e. $r(r-1) \equiv 0 \mod k$.

(2.5) Let $(P,\mathfrak{L})$ be an incidence space of type $S(2,k,v)$ and $\mathfrak{T}$ the set of all subspaces of $(P,\mathfrak{L})$. Then

a) For each $T \in \mathfrak{T}$, $(T,\mathfrak{L}(T))$ is of type $S(2,k,|T|)$ and so by (2.4) the function $r_T: T \to \mathbb{N}_0;\ p \to |\mathfrak{L}(p,T)|$ is constant with $r_T(r_T-1) \equiv 0 \mod k$.

b) If $\dim T \geq 2$ then $|T| = 1 + r_T(k-1)$, hence $r_T \geq k$ in particular, $r_E \geq k$ and $|E| = 1 + r_E(k-1) \geq 1 + k(k-1)$ for each $E \in \mathfrak{E}$. A plane $E \in \mathfrak{E}$ is projective iff $r_E = k$ and affine iff $r_E = k+1$.

c) If $E, T \in \mathfrak{T}$, $E \subset T$ and $E \neq T$ then $r_T \geq |E|$ and $|T| - |E| = 1 + |E|(k-2) + (r_T - |E|)(k-1)$.

d) Let $E \in \mathfrak{E}$ with $r_E < 1 + k(k-1)$; then $E \in \mathfrak{E}_m$.

e) For $v = k^2$, $(P,\mathfrak{L})$ is an affine plane, and for $v = k^2 - k + 1$, $(P,\mathfrak{L})$ is a projective plane.

Proof. c) Let $t \in T\setminus E$. Then for $x,y \in E$ with $x \neq y$, $\overline{t,x} \neq \overline{t,y}$, hence $r_T \geq |E|$.
d) By c), if the plane E contains a proper subplane F then $r_E \geq |F|$ and by b), $|F| \geq 1 + k(k-1)$.

3. Webs, Chain and Circular Structures

Let $r \in \mathbb{N}$. An incidence structure $(P,\mathfrak{G})$ is called an **r-web** if the functions $k^{(2)}$ and $r^{(1)}$ are constant with $k^{(2)} = 1$, $r^{(1)} = r$, if $k^{(1)} \geq 2$ and if the Euclidean parallelaxiom

(P) $\forall G \in \mathfrak{G}\ \ \forall x \in P\setminus G\ \ \exists_1 H \in \mathfrak{G}$ with $x \in H$ and $H \cap G = \emptyset$

is valid. Then, if we define as usual the parallelism $\parallel$ on the set of blocks by "$A \parallel B :\Leftrightarrow A = B$ or $A \cap B = \emptyset$", $\parallel$ is an equivalence relation on $\mathfrak{G}$. The reflexivity and the symmetry are clear by the definition of $\parallel$, and the transitivity follows from (**P**). There are exactly r parallel classes, $\mathfrak{G}_1, \mathfrak{G}_2, \ldots, \mathfrak{G}_r$ and $\mathfrak{G} = \mathfrak{G}_1 \cup \ldots \cup \mathfrak{G}_r$ is a disjoint decomposition of the set $\mathfrak{G}$ of blocks. To each $x \in P$

and to each $i \in \{1,2,\dots,r\}$ there belongs exactly one block $[x]_i \in \mathfrak{G}_i$ with $x \in [x]_i$. Therefore we can introduce the following symmetric binary relation $\frown$ on the set P of points which we call **circular:**

For $x,y \in P$ let $x \frown y :\Leftrightarrow \forall i \in \{1,2,\dots,r\}: [x]_i \neq [y]_i$.

(3.1) Let $(P, \mathfrak{G} = \mathfrak{G}_1 \cup \dots \cup \mathfrak{G}_r)$ be an r-web with $r > 1$. Then $r-1 \leq k := |X| = |Y| = |\mathfrak{G}_i|$ for all $X, Y \in \mathfrak{G}$ and all $i \in \{1,2,\dots,r\}$; if $(P,\mathfrak{B})$ is finite then $v = k^2$, $b = r \cdot k$ and $(P,\mathfrak{G})$ is an affine plane (of order k) if and only if $r = k+1$.

An incidence structure $(P, \mathfrak{B} = \mathfrak{G} \cup \mathfrak{C})$ is an **i-chain structure,** if $\mathfrak{G} \cap \mathfrak{C} = \emptyset$, if $(P,\mathfrak{G})$ is an i-web and if the condition

(C) $\forall G \in \mathfrak{G}, \quad \forall C \in \mathfrak{C}: \; |G \cap C| = 1$

is valid. The blocks of $\mathfrak{C}$ are called **chains**, and of $\mathfrak{G}$ **generators.**

(3.2) Let $(P, \mathfrak{G}_1 \cup \dots \cup \mathfrak{G}_r)$ be an r-web with $r > 1$, let $i \in \mathbb{N}$ with $1 \leq i < r$ and let $\mathfrak{G} := \mathfrak{G}_1 \cup \mathfrak{G}_2 \cup \dots \cup \mathfrak{G}_i$, $\mathfrak{C} := \mathfrak{G}_{i+1} \cup \dots \cup \mathfrak{G}_r$; then $(P, \mathfrak{G} \cup \mathfrak{C})$ is an i-chain structure.

From **(3.1)** we obtain:

(3.3) Let $(P, \mathfrak{G} \cup \mathfrak{C})$ be an i-chain structure, let $\mathfrak{G} = \mathfrak{G}_1 \cup \mathfrak{G}_2 \cup \dots \cup \mathfrak{G}_i$, and let $v,b,k,r = i$ be the parameters of the i-web $(P,\mathfrak{G})$ if $(P,\mathfrak{G})$ is finite.

a) If $\mathfrak{K} := \mathfrak{C} \cup \mathfrak{G}_2 \cup \mathfrak{G}_3 \cup \dots \cup \mathfrak{G}_i$ then $(P, \mathfrak{G}_1 \cup \mathfrak{K})$ is a 1-chain structure.

b) If $i > 1$ then $v = k^2$, $b = i \cdot k$ and $|X| = k$ for all $X \in \mathfrak{G} \cup \mathfrak{C}$.

c) If $i = 1$ then $v = b \cdot k$ and $|X| = b$ for all $X \in \mathfrak{C}$.

Derivation of i-chain-structures for i = 1 and i = 2. Let $(P, \mathfrak{G} \cup \mathfrak{C})$ be an i-chain structure with $1 \leq i \leq 2$. For a fixed point $q \in P$ let $\mathfrak{C}^q := \{C \setminus \{q\} \mid C \in \mathfrak{C},\, q \in C\}$. If $i = 1$ we set $P^q := P \setminus [q]$ and $\mathfrak{G}^q = \mathfrak{G} \setminus \{[q]\}$ and if $i = 2$ we put $P^q := P \setminus ([q]_1 \cup [q]_2)$, and $\mathfrak{G}^q := \left\{X \cap P^q \mid X \in \mathfrak{G} \setminus \{[q]_1, [q]_2\}\right\}$. Then $(P,\mathfrak{G},\mathfrak{C})^q := (P^q, \mathfrak{G}^q \cup \mathfrak{C}^q)$ is called the **derivation** in the point q. Clearly if $P^q \neq \emptyset$, then $(P,\mathfrak{G},\mathfrak{C})^q$ is again an i-chain- structure.

Particular i-chain-structures. An i-chain-structure $(P, \mathfrak{G} \cup \mathfrak{C})$ is called an **i-structure** if

I_2 $\forall x,y \in P, \; x \frown y \; \exists_1 C \in \mathfrak{C}: \; x,y \in C$,

a **circular structure** if

I_3 $\forall x,y,z \in P, \quad x \frown y \frown z \frown x \ \exists_1 C \in \mathfrak{C}: \ x,y,z \in C.$

A finite i-structure $(P, \mathfrak{G} \cup \mathfrak{C})$ with $i \geq 2$ is an affine plane if we consider the set $\mathfrak{G} \cup \mathfrak{C}$ of all generators and all chains as lines. For $i = 2$ a circular structure $(P, \mathfrak{G} \cup \mathfrak{C})$ is a **hyperbola structure**, and a **Minkowski plane**, if for any point $q \in P$ the derivation $(P, \mathfrak{G} \cup \mathfrak{C})^q$ is an affine plane. Any Minkowski plane is a hyperbola structure and in the finite case, a hyperbola structure is also a Minkowski plane.

A circular structure $(P, \mathfrak{G} \cup \mathfrak{C})$ with $i = 1$ we will call a **Laguerre-structure**; if for each point $q \in P$ the derivation $(P, \mathfrak{G} \cup \mathfrak{C})^q$ is an affine plane then $(P, \mathfrak{G} \cup \mathfrak{C})$ is a **Laguerre plane.** If we allow $\mathfrak{G} = \emptyset$, hence $x \frown y \Leftrightarrow x \neq y$, then a circular structure $(P, \mathfrak{C})$ is called a **Möbius structure,** and a **Möbius plane,** if for each $q \in P$ the derivation $(P, \mathfrak{C})^q$ is an affine plane.

By the **touching axiom** we understand the following statement:

T $\forall A \in \mathfrak{C}, \ \forall a \in A, \ \forall b \in P \backslash A$ with $a \frown b \ \exists_1 B \in \mathfrak{C}$ such that $b \in B$ and $B \cap A = \{a\}$.

(3.4) If for an Möbius structure $(P, \mathfrak{C})$ the touching axiom is fulfilled then the function $k = |X|$ for $X \in \mathfrak{C}$ is constant, $(P, \mathfrak{C})$ is a Steiner system of type $S(3,k,k)$ or $S(3,k^2-2k+2)$ and $(P, \mathfrak{C})$ is a Möbius plane if $|\mathfrak{C}| > 1$.

Proof. Let $|\mathfrak{C}| > 1$ and let $A, B \in \mathfrak{C}$ with $A \neq B$. Then there are $a \in A \backslash B$ and $b \in B \backslash A$. Because of I_3 and **T** for each $x \in A$ there is exactly one $X \in [a,b]$ with $X \cap A = \{a,x\}$. Hence $|A| \leq \big|[a,b]\big|$. By **T** there is exactly one $C \in [a,b]$ with $C \cap A = \{a\}$, hence for $Y \in [a,b] \backslash \{C\}$, $\big|Y \cap A\big| = 2$ and so $|A| = \big|[a,b]\big| = |B|$.

Now let $0 \in A \backslash \{a\}$, and $D \in [a,0,b]$. Then by **T** to each $z \in P \backslash A$ there is exactly one $Z \in [a,z]$ with $Z \cap A = \{a\}$. Since $D \cap A = \{a,0\}$, we have $\big|Z \cap D \backslash \{a\}\big| = 1$. Therefore $v-1 = (k-1)^2$.

(3.5) Let $(P, \mathfrak{G} \cup \mathfrak{C})$ be a 1-structure and let $k := |X|$ for $X \in \mathfrak{C}$ and $q := |Y|$ for $Y \in \mathfrak{G}$. Then $k \leq q + 1$. If there is a parallelism $\|'$ for $(P, \mathfrak{C})$ then $k \leq q$.

Proof. Let $A \in \mathfrak{C}$, $p \in P \backslash A$, $p' := [p] \cap A$ and $K \in \mathfrak{G} \backslash \{[p]\}$. Since $(P, \mathfrak{G} \cup \mathfrak{C})$ is a 1-structure there is for each $x \in A \backslash \{p'\}$ exactly one chain $\overline{x,p} \in \mathfrak{K}$ with

$x,p \in \overline{x,p}$ and for $x,y \in A\setminus\{p'\}$ with $x \neq y$ we have $\overline{x,p} \neq \overline{y,p}$, and so $x' := \overline{x,p} \cap K \neq y' := \overline{y,p} \cap K$. Consequently $': \begin{cases} A\setminus\{p'\} \to K \\ x \to x' := \overline{x,p} \cap K \end{cases}$ is an injection, and so $k-1 = |A| \le |K| = q$. If $\|'$ is a parallelism of $(P,\mathfrak{C})$ then $A \cap (p \|' A) = \emptyset$, but $|(p \|' A) \cap K| = 1$, hence $|A| \le |K|$.

Remarks. Let $(P,\mathfrak{G},\mathfrak{C})$ be a 1-structure with the parameters k and q, let $\mathfrak{L} := \mathfrak{G} \cup \mathfrak{C}$, $\overline{P} := P \dot{\cup} \{\infty\}$, $\overline{\mathfrak{G}} := \{\overline{X} := X \cup \{\infty\} \mid X \in \mathfrak{G}\}$ and $\overline{\mathfrak{L}} := \overline{\mathfrak{G}} \cup \mathfrak{C}$.

1. If $k = q+1$ then $(\overline{P},\overline{\mathfrak{L}})$ is a projective plane of order q. Vice versa, if $(\overline{P},\overline{\mathfrak{L}})$ is a projective plane of order q, $\infty \in \overline{P}$ a distinct element, $\mathfrak{G} := \{X\setminus\{\infty\} \mid \infty \in X \in \overline{\mathfrak{L}}\}$ and $\mathfrak{C} := \{X \in \overline{\mathfrak{L}} \mid \infty \notin X\}$, then $(\overline{P}\setminus\{\infty\}, \mathfrak{G}, \mathfrak{C})$ is a 1-structure with $k = q+1$.

2. If $k = q$ then $(P,\mathfrak{L})$ is an affine plane. Vice versa if $(P,\mathfrak{L})$ is an affine plane of order q, $A \in \mathfrak{L}$ a distinct line, $\mathfrak{G} := \{X \in \mathfrak{L} \mid X \parallel A\}$ and $\mathfrak{C} := \{X \in \mathfrak{L} \mid X \nparallel A\}$ then $(P,\mathfrak{G},\mathfrak{C})$ is a 1-structure with $q = k$.

3. Till now the order of each known finite projective or affine plane is the power of a prime.

4. To each prime power q there are projective and affine planes of order q. Therefore to each prime power q, to each natural number k with $2 < k \le q+1$ there is a 1-structure with the parameters k and q: We may assume $k < q$.

Let K be the field with q elements, let $(\overline{P},\overline{\mathfrak{L}})$ be the affine coordinate plane over the field K, let $A \in \overline{\mathfrak{L}}$ and let $\mathfrak{G}$ be a set consisting of k lines parallel to A. We set $P := \cup \{G \mid G \in \mathfrak{G}\}$ and $\mathfrak{C} := \{X \cap P \mid X \in \overline{\mathfrak{L}}$ with $X \nparallel A\}$; then $(P,\mathfrak{G},\mathfrak{C})$ is a 1-structure with the parameters k and q.

Problem **(3.1)** If $q \in \mathbb{N}$ is not the power of a prime, to which values $k \in \mathbb{N}$ with $2 < k < q$ we can find 1-structures $(P,\mathfrak{G},\mathfrak{C})$ with the parameters k and q?

4. Connections Between Incidence Structures, Chain Structures, Incidence Matrices and Codes

Let K be a finite set called **alphabet**, consisting of q symbols with $q \ge 2$, let $n \in \mathbb{N}$ and let $V := K^n := \{\mathfrak{x} = (x_1,x_2,\ldots,x_n) \mid x_i \in K\}$; then any subset $\mathfrak{C} \subset V$ is called a **q-nary code of length n.** If $q = p^d$ is the power of prime p then the alphabet K will be always identified with the finite field of order p^d and V with the n-dimensional vector space over K. Then $\mathfrak{C}$ is called a **linear code** if $\mathfrak{C}$ is a vector subspace of (V,K). For $q = 2$, hence $K = \mathbb{Z}_2$, $\mathfrak{C}$ is called a **binary code.**

For the product set $V = K^n$ the function

$$\varrho: \begin{cases} V \times V \to \mathbb{N}_0 \\ (\mathfrak{x},\mathfrak{y}) \to \left|\{i \in \{1,2,\dots,n\} \mid x_i \neq y_i\}\right| \end{cases}$$

is called the **Hamming distance** and for a code $\mathfrak{C} \subset V$, $\delta := \inf\{\varrho(\mathfrak{x},\mathfrak{y}) \mid \mathfrak{x},\mathfrak{y} \in \mathfrak{C}, \mathfrak{x} \neq \mathfrak{y}\}$ the **minimal distance** of the code $\mathfrak{C}$.

For binary codes the function $\gamma: V \to \mathbb{N}_0$; $\mathfrak{x} \to x_1 + x_2 + \dots + x_n$ is called **weight** if here the elements $0,1$ of $\mathbb{Z}_2$ are considered as real numbers. Then $\varrho(\mathfrak{x},\mathfrak{y}) = \gamma(\mathfrak{x} + \mathfrak{y})$. Also in the case that q is not the power of a prime we choose two distinct elements, which we denote by 0 and 1. Further we set $\delta_{ij} = 0$ for $i \neq j$, $\delta_{ij} = 1$ for $i = j$ and for $\lambda \in K$ let $\lambda \cdot 0 := 0 \cdot \lambda := 0$, $\lambda \cdot 1 := 1 \cdot \lambda := \lambda$. Then we set $\mathfrak{e}_i := (\delta_{i1}, \delta_{i2}, \dots, \delta_{in})$, $\lambda \cdot \mathfrak{e}_i := (\lambda \cdot \delta_{i1}, \lambda \cdot \delta_{i2}, \dots, \lambda\delta_{in})$, $P := \{\lambda \mathfrak{e}_i \mid \lambda \in K, i \in \{1,2,\dots,n\}\}$, $G_i := K\mathfrak{e}_i = \{\lambda \mathfrak{e}_i \mid \lambda \in K\}$, $\mathfrak{G} := \{G_1, G_2, \dots, G_n\}$. Finally we identify each code word $C = (c_1, c_2, \dots, c_n) \in \mathfrak{C} \subset K^n$ with the set $\{c_1\mathfrak{e}_1, c_2\mathfrak{e}_2, \dots, c_n\mathfrak{e}_n\}$. Then $(P, \mathfrak{G} \cup \mathfrak{C})$ is a 1-chain structure.

Now let us consider a finite 1-chain structure $(P, \mathfrak{G} \cup \mathfrak{C})$. Let $q := |X|$ for $X \in \mathfrak{G}$ and let $n := |C|$ for $C \in \mathfrak{C}$. Then by **(3.3)c)** $|\mathfrak{G}| = n$, $|P| = v = q \cdot n$ and if $\mathfrak{G} = \{G_1, G_2, \dots, G_n\}$ then $P = G_1 \cup G_2 \cup \dots \cup G_n$ is a disjoint decomposition. Since $|G_1| = |G_2| = \dots = |G_n|$, there are $n-1$ bijections $\pi_j: G_j \to G_1$, $j \in \{2, \dots, n\}$.

If $\mathfrak{C}$ contains a chain C which defines a parallel class in $\mathfrak{C}$, i.e.

(P') $\forall\, x \in P \setminus C\ \exists_1\, D \in \mathfrak{C}: x \in D$ and $D \cap C = \emptyset$: we set $(x \parallel C) := D$ if $x \notin C$ and $(x \parallel C) = C$ if $x \in C$

is valid, then for π_j we can choose the "parallel projections" $\pi_j: G_j \to G_1$; $x \to (x \parallel C) \cap G_1$. We identify the points of the generator G_1 with an alphabet K consisting of q symbols. Since $q \geq 2$, we denote two of the letters with the symbols 0 and 1. Then the "coordinate map" $\Psi: P \to \mathbb{Z}_n \times K$; $x \to (j, \pi_j(x))$ for $[x] = G_j$ is a bijection. Let V be the set of all words of length n and with letters of the alphabet K, i.e. $|V| = q$. Then

$$\overline{\Psi}: \mathfrak{C} \to V;\ X \to (\pi_1(X \cap G_1), \pi_2(X \cap G_2), \ldots, \pi_n(X \cap G_n))$$

is an injection, and so $\overline{\Psi}(\mathfrak{C})$ can be considered as a q-nary code of length n.

Remark. For $n \in \mathbb{N}$ let $\mathbb{Z}_n := \{1,2,\ldots,n\}$. Then a q-nary code $\mathfrak{C}$ of length n is nothing else as a set of functions $C: \mathbb{Z}_n \to K$.

Any finite incidence structure $(P,\mathfrak{B})$ can be described by an **incidence matrix** $C = (c_{ij}) \in \mathfrak{M}_{b\times v}(\mathbb{Z}_2)$. For this purpose we numerate the points of P and the blocks of $\mathfrak{B}$, hence

$P = \{p_1, p_2, \ldots, p_v\}$ and $\mathfrak{B} = \{B_1, B_2, \ldots, B_b\}$, and we put $c_{ij} := \begin{cases} 1 \text{ if } p_j \in B_i \\ 0 \text{ if } p_j \notin B_i. \end{cases}$

But any incidence matrix $C \in \mathfrak{M}_{b\times v}(\mathbb{Z}_2)$ defines a binary code $\mathfrak{C} \subset (\mathbb{Z}_2)^v$ of length v by taking the row vectors of C as code words of $\mathfrak{C}$. On the other hand, if $\mathfrak{C} \subset (\mathbb{Z}_2)^n$ is a binary code then there corresponds an incidence structure $(P,\mathfrak{B})$ with $v = |P| = n$ and $b = |\mathfrak{B}| = |\mathfrak{C}|$: For $\mathfrak{x},\mathfrak{y} \in (\mathbb{Z}_2)^n$ let $\mathfrak{x}\cdot\mathfrak{y} := x_1y_1 + x_2y_2 + \ldots + x_ny_n \in \mathbb{N}_0$, further $P := \{\mathfrak{e}_1, \mathfrak{e}_2, \ldots, \mathfrak{e}_n\}$ and for $\mathfrak{c} \in \mathfrak{C}$ let $\overline{\mathfrak{c}} := \{\mathfrak{e}_i \in P \mid \mathfrak{c}\cdot\mathfrak{e}_i = 1\}$ and $\mathfrak{B} := \{\overline{\mathfrak{c}} \mid \mathfrak{c} \in \mathfrak{C}\}$.

Therefore we can identify finite incidence structures and binary codes. If $\mathfrak{C} \subset (\mathbb{Z}_2)^v$ is a binary code then we call the elements $\mathfrak{e}_i = (\delta_{i1}, \delta_{i2}, \ldots, \delta_{iv})$ **basic words** or **points** and we identify the code words $\mathfrak{c} \in \mathfrak{C}$ with the blocks $\overline{\mathfrak{c}}$. We have $b = |\mathfrak{C}|$, v is the number of basic words, for $\mathfrak{c} \in \mathfrak{C}$, the weight $\gamma(\mathfrak{c}) := = |\overline{\mathfrak{c}}| = k^{(1)}(\overline{c})$ equals the cardinality of the block $\overline{\mathfrak{c}}$ and for $\mathfrak{x},\mathfrak{y} \in \mathfrak{C}$ with $\mathfrak{x} \neq \mathfrak{y}$, $B_1 := \overline{\mathfrak{x}}$, $B_2 := \overline{\mathfrak{y}}$ we have

$k^{(2)}(\{B_1,B_2\}) = |\overline{\mathfrak{x}} \cap \overline{\mathfrak{y}}| = \mathfrak{x}\cdot\mathfrak{y}$, $\varrho(\mathfrak{x},\mathfrak{y}) = \gamma(\mathfrak{x}+\mathfrak{y}) = \gamma(\mathfrak{x}) + \gamma(\mathfrak{y}) - 2\mathfrak{x}\cdot\mathfrak{y} = k^{(1)}(B_1) + k^{(1)}(B_2) - 2k^{(2)}(\{B_1,B_2\})$.

In the case of a tactical configuration the weight $\gamma(\mathfrak{c}) = k$ is constant for all code words $\mathfrak{c} \in \mathfrak{C}$, hence $\varrho(\mathfrak{x},\mathfrak{y}) = 2(k - k^{(2)}(\{\overline{\mathfrak{x}},\overline{\mathfrak{y}}\}))$ for all $\mathfrak{x},\mathfrak{y} \in \mathfrak{C}$. For an incidence space $r^{(2)} = 1$ implies $k^{(2)} \leq 1$, and so $\varrho(\mathfrak{x},\mathfrak{y}) \geq \gamma(\mathfrak{x}) + \gamma(\mathfrak{y}) - 1$, and if the incidence space $\mathfrak{C}$ is even a tactical configuration, the minimal distance of $\mathfrak{C}$ is $d = 2(k-1)$.

To these incidence spaces belong the **projective planes** of order n, which have the parameters $v = 1 + n + n^2 = b$, $k = r = n+1$ and $\varrho(\mathfrak{x},\mathfrak{y}) = d = 2n$ is constant for all $\mathfrak{x},\mathfrak{y} \in \mathfrak{C}$ with $\mathfrak{x} \neq \mathfrak{y}$, and the **affine planes** of order n with the parameters $v = n^2$, $b = n^2+n$, $k = n$, $r = n+1$ and $d = 2(n-1)$; here we have code words $\mathfrak{x},\mathfrak{y} \in \mathfrak{C}$ with $\varrho(\mathfrak{x},\mathfrak{y}) = 2(n-1)$ (if $\overline{\mathfrak{x}} \cap \overline{\mathfrak{y}} \neq \emptyset$) and $\mathfrak{x},\mathfrak{y} \in \mathfrak{C}$ with $\varrho(\mathfrak{x},\mathfrak{y}) = 2n$ if $\overline{\mathfrak{x}} \cap \overline{\mathfrak{y}} = \emptyset$.

The **automorphism group** of the all code V is equal the automorphism group $\mathrm{Aut}(P,\mathfrak{G})$ of the 1-web $(P,\mathfrak{G})$. $\mathrm{Aut}(P,\mathfrak{G})$ contains the n normal subgroups $S^i_K := \{\sigma \in S_P \mid \sigma(x) = x \text{ for } x \in P\setminus E_i\}$, and so the normal subgroup $T := S^1_K \times S^2_K \times \ldots \times S^n_K$ and the group of permutations $\tau' \in S'_n$ defined by: If $\tau \in S_n$ and $\mathfrak{x} = \{x_1\mathfrak{e}_1, x_2\mathfrak{e}_2, \ldots, x_n\mathfrak{e}_n\}$ then $\tau'(\mathfrak{x}) = = \{x_1\mathfrak{e}_{\tau(1)}, x_2\mathfrak{e}_{\tau(2)}, \ldots, x_n\mathfrak{e}_{\tau(n)}\}$. $\mathrm{Aut}(P,\mathfrak{G}) = T\cdot S'_n$ is the semidirect product of the groups T and S'_n. The automorphism group $\mathrm{Aut}\,\mathfrak{C}$ of a code $\mathfrak{C} \subset V$ is the subgroup

$\mathrm{Aut}\,\mathfrak{C} := \{\alpha \in \mathrm{Aut}(P,\mathfrak{G}) \mid \forall\, C \in \mathfrak{C}: \alpha(C) \in \mathfrak{C}\}$ of $\mathrm{Aut}(P,\mathfrak{G})$.

5. Incidence Spaces with m-Lines where each Line has the Same Cardinality

In this section we consider incidence spaces $(P,\mathfrak{L})$ of type $S(2,k)$, i.e. each line $L \in \mathfrak{L}$ is incident with the same number $k \in \mathbb{N}$ of points. Then by **(2.5)** we have:

(5.1) For each finite subspace $T \in \mathfrak{T}$ the function r_T is a constant with $r_T(r_T-1) \equiv 0 \mod k$.

For a line $A \in \mathfrak{L}$ the set $\mathfrak{E}(A)$ resp. $\mathfrak{E}_s(A)$ is a book resp. a small book, if for all $E, F \in \mathfrak{E}(A)$ resp. $E, F \in \mathfrak{E}_s(A)$ we have $|E| = |F|$. In this case we call $\mathfrak{E}(A)$ resp. $\mathfrak{E}_s(A)$ a **uniform** resp. a **small uniform book.** Let $\mathfrak{L}_{su} := \{A \in \mathfrak{L} \mid \mathfrak{E}_s(A) \text{ is uniform}\}$ be the set of all lines which are spines of small uniform books. Between $\mathfrak{L}_{su}$ and the sets $\mathfrak{L}_m$ of m-lines, $m \in \mathbb{N}_0$ we have here the connections:

(5.2) Let $(P,\mathfrak{L})$ be an incidence space of type $S(2,k)$ and if $\mathfrak{L}_m \neq \emptyset$ let $A \in \mathfrak{L}_m$.

a) If $\mathfrak{L}_m \neq \emptyset$ then $m(m-1) \equiv 0 \mod k$ and $\mathfrak{L}_{su} = \cup\,\{\mathfrak{L}_m \mid m \in \mathbb{N}_0: m(m-1) \equiv 0 \mod k\}$.

b) $\forall\, E \in \mathfrak{E}_s(A)$: $r_E = m+k$ and $|E| = 1+(m+k)(k-1)$.

c) $(P\setminus A, \{E\setminus A \mid E \in \mathfrak{E}_s(A)\}, \subset)$ is an incidence structure of type $S(1,(m+k-1)(k-1))$.

d) If $(P,\mathfrak{L})$ is finite and $e_A := |\mathfrak{E}_s(A)|$ then $e_A = 1$ or $e_A \geq k$, $v-k = e_A(|E|-k) = e_A(m+k-1)(k-1)$ and $r-1 = e_A(r_E-1) = e_A(m+k-1)$.

Proof. 1. Let $L \in \mathfrak{L}_{su}$. For each $x \in P \setminus L$, $E_x := \overline{\{x\} \cup L}$ is a plane with $E_x \in \mathfrak{E}_s(L)$ and $\mathfrak{L}(x, E_x) = (x \parallel L) \,\dot{\cup}\, \{\overline{x,y} \mid y \in L\}$ is a disjoint decomposition. From $r_{E_x} = |\mathfrak{L}(x, E_x)| = |(x \parallel L)| + |\{\overline{x,y} \mid y \in L\}| = |(x \parallel L)| + |L|$ and $|L| = k$ we obtain $|(x \parallel L)| = r_{E_x} - k$. Since $L \in \mathfrak{L}_{su}$, $|E_x|$ is constant, and since by (**2.4**) $|E_x| = 1 + r_{E_x}(k-1)$, also r_{E_x} is constant for all $x \in P \setminus L$. This implies that $m := |(x \parallel L)| = r_{E_x} - k$ is constant for all $x \in P \setminus L$, hence L is an m-line. This gives us $\mathfrak{L}_{su} \subset \bigcup_{m \in \mathbb{N}_0} \mathfrak{L}_m$.

2. Now suppose that $A \in \mathfrak{L}_m$ is an m-line. Then for each $x \in P \setminus A$, $E_x := \overline{\{x\} \cup A} \in \mathfrak{E}_s(A)$, $r_{E_x} = m + k$, and so $|E_x| = 1 + (m+k)(k-1)$. Thus the cardinality for all planes $E \in \mathfrak{E}_s(A)$ is equal $1 + (m+k)(k-1)$, i.e. $A \in \mathfrak{L}_{su}$.

Since by (**2.5**)**a**) $0 \equiv r_E(r_E - 1) = (m+k)(m+k-1) \equiv m(m-1) \mod k$, **1.** and **2.** yield **a**) and **b**).

c) is a consequence of $A \in \mathfrak{L}_m \subset \mathfrak{L}_{su}$ and **b**).

d) Let $E \in \mathfrak{E}_s(A)$, $b \in E \setminus A$, $c \in P \setminus E$ and $L := \overline{b,c}$ if $E \neq P$. Since $L \cap E = \{b\}$ we have $\overline{x \cup A} \neq \overline{y \cup A}$ for all $x, y \in L$ with $x \neq y$. This shows $e_A = |\mathfrak{E}_s(A)| \geq |L| = k$ for $E \neq P$. **c**) implies the formulas of **d**).

(**5.3**) For a point $p \in P$ each of the statements **Ia**) and **Ib**) of (**1.3**) is equivalent to the following statement

Ic) $\forall E, F \in \mathfrak{E}_s(p)$, $E \neq F$: $|E \cap F| \leq k$

and also **IIa**), **IIb**), **IIc**) are equivalent:

IIa) $[p] \subset \mathfrak{L}_{su}$

IIb) $\forall E, F \in \mathfrak{E}_s(p)$: $|E| = |F|$

IIc) $([p], \mathfrak{E}_s(p))$ is of type $S(2, r_E)$ with $r_E = \frac{|E| - 1}{k - 1}$.

Proof. **Ic**) $\Rightarrow$ **Ia**). Let $A \in [p]$ and $E, F \in \mathfrak{E}_s(A)$ with $E \neq F$. Then $E, F \in \mathfrak{E}_s(p)$ and $k \doteq |A| \leq |E \cap F| \leq k$ by **Ic**). Hence $A = E \cap F$, i.e. $A \in \mathfrak{L}_s$.

Ia) $\Rightarrow$ **Ic**). Let $E, F \in \mathfrak{E}_s(p)$ with $E \neq F$. Then, since by (**1.3**) $([p], \mathfrak{E}_s(p))$ is an incidence space, we have either $E \cap F = \{p\}$ or $L := E \cap F \in \mathfrak{L}$, hence $|E \cap F| = 1$ or $|E \cap F| = k$.

Suppose **IIa**) holds; then also **Ia**), hence **Ib**). Let $E, F \in \mathfrak{E}_s(p)$ with $E \neq F$. If

$|E \cap F| \geq 2$ then $A := E \cap F \in [p]$ and $E, F \in \mathfrak{E}_S(A)$ by **Ib**), hence $|E| = |F|$ by **IIa**). If $E \cap F = \{p\}$ let $a \in E \setminus \{p\}$, $b \in F \setminus \{p\}$ and $D := \{\overline{p,a,b}\}$. Then $D \in \mathfrak{E}_S(p)$, $|D \cap E|, |D \cap F| \geq 2$, and so $|E| = |D| = |F|$. Thus **IIa**) $\Longleftrightarrow$ **IIb**).

Each plane $E \in \mathfrak{E}_S(p)$ is incident with exactly r_E lines of $[p]$. If we assume **IIb**) then also **Ib**) is valid and r_E is constant for all $E \in \mathfrak{E}_S(p)$, thus $\big([p], \mathfrak{E}_S(p), \subset\big)$ is of type $S(2, r_E)$. Thus **IIb**) $\Longleftrightarrow$ **IIc**).

Let P_I resp. P_{II} be the set of all points $p \in P$ for which one of the equivalent conditions I resp. II holds. If $p \in P_{II}$ then by **IIb**) and (**5.2**), $|E|$ and r_E are constant for all $E \in \mathfrak{E}_S(p)$ and e_A is constant for all $A \in [p]$ if $(P, \mathfrak{L})$ is finite.

Since $r_E = |\mathfrak{L}(p, E)|$ is the number of "points" of $[p]$ which are incident with a block of $\mathfrak{E}_S(p)$, we have $e_A(e_A - 1) \equiv 0 \mod r_E$ by (**2.4**). This takes care that the number $\frac{r \cdot e_A}{r_E}$ is an integer which equals $e_p := |\mathfrak{E}_S(p)|$. Hence

(**5.4**) If $(P, \mathfrak{L})$ is finite then for $p, q \in P_{II}$ we have:

a) The functions r_E and $|E|$ are constant for all $E \in \mathfrak{E}_S(p)$ and e_A is constant for all $A \in [p]$ and $e_A(e_A - 1) \equiv 0 \mod r_E$.

b) $r_E(p) = r_E(q)$ and $e_A(p) = e_A(q)$.

Proof. b) If $p \neq q$ let $A := \overline{p,q}$; then by definition $e_A(p) = |\mathfrak{E}_S(A)| = e_A(q)$ and $r_E(p) = r_E(q)$ for each plane $E \in \mathfrak{E}_S(A)$.

Because of (**5.4**) we have a classification of all incidence spaces $(P, \mathfrak{L})$ of type $S(2, k, v)$ with $P_{II} \neq \emptyset$. Let $\beta, \gamma, \delta, \varepsilon \in \mathbb{N}_0$ with $\varepsilon \neq 0$, $\gamma < k$, $\delta < \gamma + \varepsilon k$. An incidence space $(P, \mathfrak{L})$ of type $S(2, k, v)$ with $P_{II} \neq \emptyset$ is called of **type** $[\gamma, \varepsilon; \delta, \beta]$ if for $p \in P_{II}$, $A \in [p]$, $E \in \mathfrak{E}_S(p)$ we have $r_E = \gamma + \varepsilon k$ and $e_A = \delta + \beta r_E = \delta + \beta(\gamma + \varepsilon k)$.

(**5.5**) **Theorem.** Let $(P, \mathfrak{L})$ be of type $[\gamma, \varepsilon; \delta, \beta]$. Then the parameters of $(P, \mathfrak{L})$ have the following values:

$r_E = \gamma + \varepsilon k$ and $|E| = 1 + (\gamma + \varepsilon k)(k-1)$ for all $E \in \cup\{\mathfrak{E}_S(p) \mid p \in P_{II}\}$ and
$\gamma(\gamma - 1) \equiv 0 \mod k$,

$e_A = \delta + \beta r_E = \delta + \beta(\gamma + \varepsilon k)$ for all $A \in \cup \big\{[p] \mid p \in P_{II}\big\}$ and $\delta(\delta - 1) \equiv 0 \mod r_E$,

$r = 1 + e_A(r_E - 1) = 1 + (\delta + \beta(\gamma + \varepsilon k)(\gamma - 1 + \varepsilon k),$

$e_p = \frac{r \cdot e_A}{r_E}$ for all $p \in P_{II}$,

$v = k + (|E| - k)e_A = k + (\gamma + \varepsilon k - 1)(k - 1)(\delta + \beta(\gamma + \varepsilon k)),$

$e_p = \beta \cdot r = \beta \cdot [1 + \beta(\gamma + \varepsilon k)(\gamma - 1 + \varepsilon k)]$, if $\delta = 0$,

$e_p = e_A(e_A - \beta) = (1 + \beta(\gamma + \varepsilon k))(1 + \beta(\gamma - 1 + \varepsilon k))$ if $\delta = 1$.

Proof. By $e_p \cdot r_E = r \cdot e_A = r(\delta + \beta r_E)$ we have $e_p = \beta r$ if $\delta = 0$ and since $r = e_A r_E - (e_A - 1)$ we have $r = e_A r_E - \beta r_E$ for $\delta = 1$, hence $e_p = e_A(e_A - \beta)$. By (**5.3**) we have:

(**5.6**) **a**) For $L \in \mathfrak{L}$, "$L \cap P_I \neq \emptyset \iff L \in \mathfrak{L}_S$" and "$L \cap P_{II} \neq \emptyset \iff L \in \mathfrak{L}_{su}$".

b) If $\mathfrak{E}_S(p) \subset \mathfrak{E}_m$ then $p \in P_I$.

(**5.7**) the following statements are equivalent:

a) $P = P_I$

b) $\mathfrak{E} = \mathfrak{E}_m$

c) $\mathfrak{L} = \mathfrak{L}_S$

d) Each $A \in \mathfrak{L}$ is the spine of a book.

Proof. By (**5.5**) we have **b**) $\Rightarrow$ **a**) $\iff$ **c**). Now let us assume **c**), let $E = \{\overline{a, b, c}\} \in \mathfrak{E}$ and let $d \in E \setminus \overline{a, b}$. Then $F := \{\overline{a, b, d}\} \in \mathfrak{E}$, $E, F \in \mathfrak{E}_S(\overline{a, b})$ and $F \subset E$. Consequently $E = F$ since $\overline{a, b} \in \mathfrak{L}_S$. Thus $E \in \mathfrak{E}_m$, i.e. $\mathfrak{E} = \mathfrak{E}_m$. Finally **b**) implies **d**) and **d**) implies **c**).

From (**5.3**) we also obtain:

(**5.8**) The following statements **a**) to **d**) are equivalent and if $v = |P| \in \mathbb{N}$ then also **e**) is equivalent with **a**).

a) $P = P_{II}$,

b) $\forall E, F \in \mathfrak{E}$: $|E| = |F|$ $\quad (\iff r_E = r_F \Rightarrow \mathfrak{E} = \mathfrak{E}_m)$,

c) $\mathfrak{L} = \mathfrak{L}_{su}$,

d) $\mathfrak{L} = \mathfrak{L}_u$,

e) $(P, \mathfrak{E})$ is an incidence structure of type $T(2, e_A, |E|, v)$, hence $v \cdot e_p = |\mathfrak{E}| \cdot |E|$ implying $v \cdot e_p \equiv 0 \mod |E|$.

Remarks on incidence spaces $(P,\mathfrak{L})$ of type $S(2,k,v)$.

1. If $\mathfrak{L}_0 \neq \emptyset$ and $A \in \mathfrak{L}_0$ then $\mathfrak{E}_S(A)$ consists of projective planes of order $n := k-1$ and $v = n+1+e_A \cdot n^2$ with $e_A = 1$ or $e_A \geq n+1$.

2. If $\mathfrak{L}_1 \neq \emptyset$ and $A \in \mathfrak{L}_1$ then $\mathfrak{E}_S(A)$ consists of affine planes of order k and $v = k + e_A \cdot k(k-1)$ with $e_A = 1$ or $e_A \geq k$.

3. If $\mathfrak{L}_0 \neq \emptyset$ and $\mathfrak{L}_1 \neq \emptyset$, let $A \in \mathfrak{L}_0$ and $B \in \mathfrak{L}_1$. Then $e_A(k-1) = e_B \cdot k$. If k is the power of a prime (which we are allowed to assume since all finite affine planes, we know so far, have a prime power as their order), then $e_A = k$ and $e_B = k-1$, hence $v = k + k^2(k-1) = k^3 - k^2 + k$ are the smallest possible values.

4. If k is the power of a prime then by (**2.5**) we have either $r \equiv 0 \mod k$ or $r \equiv 1 \mod k$, and therefore $\mathfrak{T} = \mathfrak{T}_0 \dot{\cup} \mathfrak{T}_1$ with $\mathfrak{T}_0 := \{T \in \mathfrak{T} \mid r_T \equiv 0 \mod k\}$, $\mathfrak{T}_1 := \{T \in \mathfrak{T} \mid r_T \equiv 1 \mod k\}$. Further, if $\mathfrak{L}_m \neq \emptyset$, then $m \equiv 0 \mod k$ or $m \equiv 1 \mod k$.

5. Incidence spaces $(P,\mathfrak{L})$ of type $[0,1;\ \delta,\beta]$ resp.($[1,1;\ \delta,\beta]$) are characterized by $P_{II} \neq \emptyset$ and the property: If E is a plane generated by three non-collinear points, one of which is in P_{II}, then E is a projective resp. affine plane of order $n := k-1$ resp. k.

a) If $(P,\mathfrak{L})$ is of type $[0,1;\ \delta,\beta]$ and $n := k-1$ then $r_E = 1+n$, $|E| = 1+n+n^2$, $e_A = \delta + \beta(1+n)$, $r = 1+n(\delta+\beta(1+n))$, $v = 1+n+n^2(\delta+\beta(1+n))$, and $\delta(\delta-1) \equiv 0 \mod (1+n)$.

b) If $(P,\mathfrak{L})$ is of type $[1,1;\ \delta,\beta]$ then $r_E = 1+k$, $|E| = k^2$, $e_A = \delta+\beta(1+k)$, $r = 1+k(\delta+\beta(1+k))$, $v = k[1+(k-1)\cdot(\delta+\beta(1+k))]$, and $\delta(\delta-1) \equiv 0 \mod 1+k$.

6. For incidence spaces $(P,\mathfrak{L})$ with $P_{II} \neq \emptyset$, where for one $p \in P_{II}$ the bundle space $([p], \mathfrak{E}_S(p))$ is a projective space of order n and dimension d, we have $r_E = n+1$, $|E| = 1+(n+1)(k-1)$ for all $E \in \cup\{\mathfrak{E}_S(q) \mid q \in P_{II}\}$, $e_A = 1+n+\ldots+n^{d-1}$, for all $A \in \cup\{[q] \mid q \in P_{II}\}$, $r = 1+ne_A$, $v = 1+(1+n+\ldots+n^d)(k-1)$ and $(P,\mathfrak{L})$ is of type $[\gamma,\varepsilon;\ 0,\beta]$ with $\beta = 1+n^2+n^4+\ldots+n^{2m}$ if the dimension $d = 2m+2$ is even, and $(P,\mathfrak{L})$ is of type $[\gamma,\varepsilon;\ 1,\beta]$ with $\beta = n(1+n^2+n^4+\ldots+n^{2m})$ if $d = 2m+3$ is odd. If furthermore $(P,\mathfrak{L})$ is of type $[0,1;0,\beta]$ or $[0,1;1,\beta]$ then $n = r_E - 1 = k-1$, hence $k = 1+n$, and $(P,\mathfrak{L})$ has the same parameters as the $(d+1)$-dimensional projective space of order $k-1$, where in the first case the dimension of $(P,\mathfrak{L})$ is odd in the second even. If $(P,\mathfrak{L})$ is of type $[1,1;\ \delta,\beta]$ then $(P,\mathfrak{L})$ has the same parameters as the $(d+1)$-dimensional affine space of order k and again $\delta = 0$ if $d+1$ is odd and $\delta = 1$ if $d+1$ is even.

7. If for one $p \in P_{II}$ the bundle space $([p]), \mathfrak{E}_S(p))$ is an affine space of order n and dimension d then: $n := r_E$, $|E| = 1+n(k-1)$ for all $E \in \cup \{\mathfrak{E}_S(q) \mid q \in P_{II}\}$, $e_A = 1+n+\ldots+n^{d-1}$ for all $A \in \cup \{[q] \mid q \in P_{II}\}$, $r = n^d$, $v = 1+n^d(k-1)$.

6. Incidence Spaces where Each Plane has the Same Cardinality

In this section let $(P,\mathfrak{L})$ be an incidence space of type $S(2,k)$ such that $|E| = |F|$ for all $E, F \in \mathfrak{E}$. Besides the formulas of Theorem (**5.5**), we have by (**5.8**):

(*) $v \cdot e_p \equiv 0 \mod |E|$.

Since $v = k + e_A(|E|-k)$ (cf. (**5.2**)**c**)), (*) is equivalent with:

(**) $k(1-e_A) \cdot e_p \equiv 0 \mod |E|$.

If $(P,\mathfrak{L})$ is of type $[1,\varepsilon;\delta,\beta]$ then $|E| = k(1+\varepsilon(k-1))$, and so (**) is equivalent with:

(**)**a**) $(1-e_A) \cdot e_p \equiv 0 \mod (1+\varepsilon(k-1))$.

By $\varepsilon(k-1) \equiv -1 \mod (1+\varepsilon(k-1))$ we have $r_E \equiv \varepsilon$, hence $e_A \equiv \delta + \beta\varepsilon$, and so $r \equiv 1 + (\delta+\beta\varepsilon)(\varepsilon-1)$.

Since $e_p = \beta \cdot r$ for $\delta = 0$ this yields:

(**6.1**) If $(P,\mathfrak{L})$ is of type $[1,\varepsilon;0,\beta]$ then $(1-\beta\varepsilon)\beta(1+\beta\varepsilon(\varepsilon-1)) \equiv 0 \mod (1+\varepsilon(k-1))$.

For $\delta = 1$, $(1-e_A) \equiv -\beta\varepsilon$ and $e_p = e_A(e_A-\beta) \equiv (1+\beta\varepsilon)(1+\beta(\varepsilon-1))$. Since $(\varepsilon, 1+\varepsilon(k-1)) = 1$ we obtain:

(**6.2**) If $(P,\mathfrak{L})$ is of type $[1,\varepsilon;1,\beta]$ then $\beta(1+\beta\varepsilon)(1+\beta(\varepsilon-1)) \equiv 0 \mod (1+\varepsilon(k-1))$.

Now let $(P,\mathfrak{L})$ be of type $[0,\varepsilon;\delta,\beta]$, hence $r_E = \varepsilon k$ and $|E| = 1+\varepsilon k(k-1)$. Then $(k,|E|) = 1$, and therefore (**) is equivalent with:

(**) **b**) $(1-e_A) \cdot e_p \equiv 0 \mod |E|$.

Since $\varepsilon k^2 \equiv \varepsilon k - 1 \mod |E|$ we have $e_A \equiv \delta + \beta\varepsilon k$, $r \equiv 1 + (\delta+\beta\varepsilon k)(\varepsilon k - 1) \equiv 1 - \delta - \beta\varepsilon + (\delta - \beta + \beta\varepsilon)\varepsilon k$. If $\delta = 0$ then $1 - e_A \equiv (1-\beta\varepsilon k)$, $e_p = \beta r \equiv \beta[1-\beta\varepsilon+(\varepsilon-1)\beta\varepsilon k]$ and by (**)**b**) we obtain:

(**6.3**) If $(P,\mathfrak{L})$ is of type $[0,\varepsilon;0,\beta]$ then
$\beta \cdot [1-\beta\varepsilon+\beta^2(\varepsilon-1)\varepsilon+(-2+\varepsilon+2\beta\varepsilon-\beta\varepsilon^2)\beta\varepsilon \cdot k] \equiv 0 \mod (1+\varepsilon k(k-1))$.

If $\delta = 1$ then $1 - e_A \equiv -\beta\varepsilon k$ and (**)b) is equivalent with $\beta\varepsilon k \cdot e_p \equiv 0 \bmod |E|$ and, since $(\varepsilon k, |E|) = 1$, with

(**)c) $\beta \cdot e_p \equiv 0 \bmod |E|$.

and we have $e_p = e_A(e_A - \beta) \equiv (1 + \beta\varepsilon k)\cdot(1 - \beta + \beta\varepsilon k) \equiv 1 - \beta - \beta^2\varepsilon + (2 - \beta + \beta\varepsilon)\beta\varepsilon k$.

This implies:

(6.4) If $(P,\mathfrak{L})$ is of type $[0,\varepsilon;\ 1,\beta]$ then $\beta\cdot[1-\beta-\beta^2\varepsilon+(2-\beta+\beta\varepsilon)\beta\varepsilon k] \equiv 0 \bmod (1+\varepsilon k(k-1))$.

We know that a projective space is an incidence space $(P,\mathfrak{L})$ with $\mathfrak{L} = \mathfrak{L}_0$, i.e. each line is a projective line. But this is equivalent with the statement that each plane E of $(P,\mathfrak{L})$ is a projective plane. By **(2.5**)**b**) the projective planes are determined by the parameters $\delta = 0$, $\varepsilon = 1$. Therefore we have:

(6.5) The incidence spaces of type $[0,1;\ \delta,\beta]$ are exactly the finite projective spaces $(P,\mathfrak{L})$ of order $n := k-1$. If d is the dimension of $(P,\mathfrak{L})$ then $\delta = 0$, $\beta = 1 + n^2 + n^4 + \ldots + n^{d-3}$ if d is odd, and $\delta = 1$, $\beta = n(1 + n^2 + \ldots + n^{d-4})$ if d is even.

Proof. For $\varepsilon = 1$ and $k = n+1$ the equations **(6.3)** and **(6.4)** reduce to

(6.3)' $\beta(\beta-1)(\beta n+\beta-1) \equiv 0 \bmod(1+n+n^2)$ and

(6.4)' $\beta[1+\beta-\beta^2+2\beta n] \equiv 0 \bmod(1+n+n^2)$.

Since $|P| = \dfrac{n^{d+1}-1}{n-1} = n+1+e_A n^2$, $e_A = \dfrac{n^{d+1}-n^2}{n^2(n-1)} = \dfrac{n^{d-1}-1}{n-1} = 1+n+\ldots+n^{d-2} = \delta+\beta r_E = \delta+\beta(1+n)$. If d is odd then e_A is divisible by $1+n$ and we have $e_A = (1+n)(1+n^2+n^4+\ldots+n^{d-3})$ hence $\delta = 0$ and $\beta = 1+n^2+n^4+\ldots+n^{d-3}$; if d is even then $e_A - 1$ is divisible by $1+n$, so $e_A - 1 = (1+n)n(1+n^2+n^4+\ldots+n^{d-4})$, i.e. $\delta=1$ and $\beta = n(1+n^2+n^4+\ldots+n^{d-4})$.

Remarks. By **(6.5)** the class $[0,1;\ \delta,\beta]$ with $P = P_{II}$ is not empty only if $\delta = 0$ and β is a number of the form $1+n^2+n^4+\ldots+n^{2m}$, or $\delta = 1$ and β has the form $n(1+n^2+n^4+\ldots+n^{2m})$.

An incidence space $(P,\mathfrak{L})$ is a pseudo-affine space if each plane E of $(P,\mathfrak{L})$ is an affine plane. In a pseudo-affine space to any line $A \in \mathfrak{L}$ and any point $p \in P$ there is exactly one line, denoted by $(p \parallel A)$ with $p \in (p \parallel A)$ and $(p \parallel A) \parallel A$

(i.e. $(p \parallel A)=A$ or $(P\parallel A) \cap A=\emptyset$ and $\overline{(p\parallel A) \cup A}$ is a plane), but the relation $\parallel$ need not be transitive. We know (cf. [12], [13], [10]) that $\parallel$ is transitive as soon as $k \geq 4$. With (**2.5**)**b**) this gives us the result:

(**6.6**) The incidence spaces of type $[1,1;\delta,\beta]$ are exactly the finite pseudo-affine spaces $(P,\mathfrak{L})$ of order k. If $k \geq 4$ then $(P,\mathfrak{L})$ is an affine space of order k. If $(P,\mathfrak{L})$ is a finite affine space of dimension d then $\delta=0$, $\beta=1+k^2+k^4+\ldots+k^{d-3}$, if d is odd, or $\delta=1$, $\beta=k(1+k^2+k^4+\ldots+k^{d-4})$, if d is even.

Proof. For $\varepsilon=1$ the equations (**6.1**) and (**6.2**) reduce to

(**6.1**)' $\beta(\beta-1)\equiv 0 \mod k$ and

(**6.2**)' $\beta(\beta+1)\equiv 0 \mod k$,

and we have $k^d-k=|P|-k=e_A\cdot(k^2-k)$. Thus $e_A=1+k+k^2+\ldots+k^{d-2}$ is divisible by $r_E=1+k$ if and only if d is odd, and then $\delta=0$ and $\beta=1+k^2+k^4+\ldots+k^{d-3}$; and e_A-1 is divisible by $1+k$ if d is even, and then $\delta=1$ and $\beta=k(1+k^2+k^4+\ldots+k^{d-4})$.

Remark. For $k=3$ there are pseudo- affine spaces $(P,\mathfrak{L})$ which are not affine spaces. All the same the planes of $\mathfrak{E}$ are the affine planes of order 3. Clearly if every subspace $T\in\mathfrak{T}$ which is generated by four non-complanar points is a 3-dimensional affine space then $(P,\mathfrak{L})$ is an affine space. Therefore a proper pseudo-affine space has more than 27 points. Also if the bundle space $([p],\mathfrak{E}(p))$ is a projective space for each $p\in P$ then $(P,\mathfrak{L})$ is an affine space. By (**6.1**)' and (**6.2**)' the cardinality $|P|$ of a pseudo affine space $(P,\mathfrak{L})$ of order 3 can only be one of the numbers $v=3(1+24\lambda)$, $v=9(3+8\lambda)$ $v=9(1+8\lambda)$, $v=3(19+24\lambda)$ with $\lambda\in\mathbb{N}_0$. But by [14], [15] we know already that the number of points of a pseudo affine space of order 3 is a power of 3. The smallest proper pseudo affine spaces have 81 elements.

In connection with the theorems (**6.5**) and (**6.6**) the following problems arise:

P 5.1 Are there any incidence spaces $(P,\mathfrak{L})$ of type $S(2,k)$ with $\mathfrak{L}=\mathfrak{L}_i$ for $i>1$?

By (**5.2**) i has to be a number of the form $i=r_E-k=\gamma+\varepsilon k-k=\gamma+(\varepsilon-1)k$ with $\gamma(\gamma-1)\equiv 0 \mod k$, $0\leq\gamma\leq k-1$ and $\varepsilon\in\mathbb{N}$.

P5.2 If such incidence spaces $(P,\mathfrak{L})$ with $i>1$ do exist which types $[\gamma,\varepsilon;\ \delta,\beta]$ are possible?

P5.3 Before tackling with these general problems one has to consider the particular case where $(P,\mathfrak{L})$ is a plane, i.e. $e_A = \delta+\beta r_E = 1$, hence where $(P,\mathfrak{L})$ is of type $[\gamma,\varepsilon;1,0]$. Thus, are there incidence spaces $(P,\mathfrak{L})$ which are exchange planes? By Theorem **(2.5)d)** we know only: If $(P,\mathfrak{L})$ is of type $[\gamma,\varepsilon;\ 1,0]$ with $\varepsilon \le k-1$ for $\gamma=0$ or $\varepsilon < k-1$ for $\gamma=0$ then $(P,\mathfrak{L})$ is an exchange plane.

7. Incidence Spaces with Connected Uniform Books

We say that a line $A \in \mathfrak{L}_{su}$ is of type $[\gamma,\varepsilon]$, $0 \le \gamma < k$, if for each plane $E \in \mathfrak{E}_s(A)$, $r_E = \gamma+\varepsilon k$ and $|E|-1 = r_E(k-1)$.

We call two spines $A,B \in \mathfrak{L}_{su}$ of small uniform books **directly connected** if there are $E \in \mathfrak{E}_s(A)$ and $F \in \mathfrak{E}_s(B)$ with $B \subset E$ and $A \subset F$ and **connected** if there is a sequence $A = X_1, X_2, \dots, X_n = B$ with $X_i \in \mathfrak{L}_{su}$ such that for all $i \in \{1,2,\dots,n-1\}$ X_i and X_{i+1} are directly connected.

(**7.1**) If two spines $A,B \in \mathfrak{L}_{su}$ are connected then A and B are of the same type and $e_A = e_B$.

Proof. Let $A \ne B$ and let $E \in \mathfrak{E}_s(A)$, $F \in \mathfrak{E}_s(B)$ with $B \subset E$ and $A \subset F$. Since A and B are spines of small books we have $E = \overline{A \cup \{x\}}$ for all $x \in E\setminus A$ in particular $E = \overline{A \cup B} = F$. This implies $r_E = r_F$ and $e_A = \dfrac{v-k}{|E|-k} = e_B$.

We recall (cf. (**5.2**)**a**)) that $\mathfrak{L}_{su} = \cup\{\mathfrak{L}_i \mid i \in N_0:\ i(i-1) \equiv 0 \bmod k\}$ is the set of all i-lines. If we assume that $\mathfrak{L}_{su}$ is connected then $\mathfrak{L}_{su} = \mathfrak{L}_j$ for $j = r_E - k$, $E \in \mathfrak{E}_s(A)$ and $A \in \mathfrak{L}_{su}$. We notice that $\mathfrak{L}_{su}$ is connected if $P_{II} \ne \emptyset$: For if $p \in P_{II}$, $A \in [p]$ and $B \in \mathfrak{L}_{su}$ with $B \ne A$, let $b \in B\setminus A$ and $C := \overline{p,b}$. If $C \ne A$ then $C \in \mathfrak{L}_{su}$ and A and C are directly connected because by $p \in A,C$, $E := \overline{A \cup C}$ is a plane such that $E \in \mathfrak{E}_s(A)$ and $E \in \mathfrak{E}_s(C)$. Further $b \in B \cap C$ implies that $F := \overline{B \cup C}$ is a plane with $F \in \mathfrak{E}_s(C)$ and $F \in \mathfrak{E}_s(B)$. $P_{II} \ne \emptyset$ implies also $\cup\,\mathfrak{L}_{su} = P$ and $e_A = \delta+\beta r_E = \delta+\beta(\gamma+\varepsilon k)$ for all $A \in \mathfrak{L}_{su}$.

By a **spread** $\mathfrak{S}$ one understands a subset of $\mathfrak{L}$ such that to each $x \in P$ there is exactly one line $X \in \mathfrak{S}$ with $x \in X$. The existence of a spread yields $v \equiv 0 \bmod k$. So let us assume $v \equiv 0 \bmod k$. Then $v - 1 = r(k-1)$ implies $r-1 \equiv 0$

mod k, and if $A \in \mathfrak{L}_{su}$, $E \in \mathfrak{E}_s(A)$ then $r-1 = e_A(r_E-1)$ and $v-k = e_A(|E|-k)$ yield $e_A(r_E-1) \equiv e_A \cdot |E| \equiv 0 \mod k$.

Let $A \in \mathfrak{L}_{su}$ be of type $[\gamma, \varepsilon]$, hence $e_A(\gamma-1) \equiv 0 \mod k$.

If $\gamma = 0$ then $e_A = \beta k$ with $\beta \in \mathbb{N}$, $r-1 = \beta k(\varepsilon k-1)$, $|E|-k = (r_E-1)(k-1)$ and so $v-k = (r-1)(k-1) = \beta k(k-1)(\varepsilon k-1)$ or $v = k(1+\beta(k-1)(\varepsilon k-1))$.

If $\gamma = 1$ then $r-1 = \varepsilon e_A k$, $v-k = \varepsilon e_A k(k-1)$ or $v = k(1+\varepsilon e_A(k-1))$.

Suppose now, that $\mathfrak{L}_{su}$ contains a spread $\mathfrak{S}$ such that any two lines of $\mathfrak{S}$ are directly connected. By (**7.1**) all lines of $\mathfrak{S}$ are of the same type $[\gamma, \varepsilon]$. For $A \in \mathfrak{S}$, $E \in \mathfrak{E}_s(A)$ and $b \in E \setminus A$ let $B \in \mathfrak{S}$ with $b \in B$. Since A, B are directly connected and $b \in E$ we have $B \subset E$. This shows us that $\mathfrak{S} \cap \mathfrak{L}(E)$ is a spread of E implying $r_E - 1 \equiv 0 \mod k$ and so $\gamma = 1$, i.e. $\mathfrak{S}$ is of type $[1, \varepsilon]$. If we put $\mathfrak{E}_{\mathfrak{S}} := \bigcup \{\mathfrak{E}_s(A) \mid A \in \mathfrak{S}\}$ then for each plane $E \in \mathfrak{E}_{\mathfrak{S}}$, $r_E = 1+\varepsilon k$, $|E| = k(1+\varepsilon(k-1))$ and so $\mathfrak{S} \cap \mathfrak{L}(E)$ contains exactly $1+\varepsilon(k-1)$ lines. Since $\mathfrak{S}$ is directly connected we have for $A, B \in \mathfrak{S}$ with $A \neq B$, $\overline{A \cup B} \in \mathfrak{E}_{\mathfrak{S}}$. Thus $(\mathfrak{S}, \mathfrak{E}_{\mathfrak{S}})$ is an incidence space of type $S(2, 1+\varepsilon(k-1))$ implying $e_A(e_A-1) \equiv 0 \mod (1+\varepsilon(k-1))$ for $A \in \mathfrak{S}$. If $e_A \neq 1$ then of course $e_A \geq k$. We have proved:

(**7.2**) Let $\mathfrak{L}_{su} \neq \emptyset$, $A \in \mathfrak{L}_{su}$ and let A be of type $[\gamma, \varepsilon]$.

a) If $v \equiv 0 \mod k$ and $\gamma = 0$ then there is a $\beta \in \mathbb{N}$ with $e_A = \beta k$ and so $v = k(1+\beta(k-1)(\varepsilon k-1))$. In particular for $[0,1]$ and $n := k-1$, $v = (1+n)(1+\beta n^2) = 1+n+\beta(n^2+n^3)$.

b) If $\gamma = 1$ then $v \equiv 0 \mod k$, $e_A = 1$ or $e_A \geq k$ and $v = k(1+\varepsilon e_A(k-1))$.

c) If $\mathfrak{L}_{su}$ contains a directly connected spread $\mathfrak{S}$ then $\gamma = 1$, $e_A(e_A-1) \equiv 0 \mod(1+\varepsilon(k-1))$ and $(\mathfrak{S}, \mathfrak{E}_{\mathfrak{S}})$ is an incidence space of type $S(2, 1+\varepsilon(k-1))$. In particular for $[1,1]$, $e_A = \delta+\beta k$ and $v = k(1+(\delta+\beta k)(k-1))$, hence $v = k(1+\beta k(k-1))$ or $v = k^2(1+\beta(k-1))$.

These considerations lead to the following problems:

P7.1 Are there incidence spaces $(P, \mathfrak{L})$ such that $\mathfrak{L}_{su} = \mathfrak{L}_0$ and $P_{II} \neq \emptyset$ which are not projective spaces?

P7.2 Are there non-projective incidence spaces $(P, \mathfrak{L})$ such that $\mathfrak{L}_0$ is connected and that $\mathfrak{L}_0$ contains a spread?

If $(P,\mathfrak{L})$ is a pseudo-affine space, i.e. $\mathfrak{L} = \mathfrak{L}_1$, and if we fix $A \in \mathfrak{L}$ then $\mathfrak{S} := \{X \in \mathfrak{L} \mid X \parallel A\}$ forms a directly connected spread.

If $(P,\mathfrak{L})$ is a projective space, hence $\mathfrak{L} = \mathfrak{L}_0$ then by **(7.2)** $\mathfrak{L}$ can contain a spread $\mathfrak{S}$ only if the dimension of $(P,\mathfrak{L})$ is odd. If $\dim(P,\mathfrak{L}) = 3$ then $(P,\mathfrak{L})$ has a finite field K as coordinate domain such that P can be identified with the set $\left\{ K^* \begin{pmatrix} a, b \\ c, d \end{pmatrix} \middle| \ a,b,c,d \in K,\ (a,b,c,d) \neq (0,0,0,0) \right\}$ of all homogeneous 2×2-matrices.

If $\mathrm{char}(K) \neq 2$, there is a $\varkappa \in K$ which is not a square. Then $A = K\begin{pmatrix} 1 & 0 \\ 0 & 1 \end{pmatrix} + K\begin{pmatrix} 0 & 1 \\ \varkappa & 0 \end{pmatrix}$ is a 2-dimensional vector subspace of the vector space $\mathfrak{M}_{22}(K)$ of all 2×2-matrices with coefficients in K and each matrix of A unequal the zero matrix has an inverse. Also $B = K\begin{pmatrix} 1 & 0 \\ 0 & 0 \end{pmatrix} + K\begin{pmatrix} 0 & 0 \\ 1 & 0 \end{pmatrix}$ is a 2-dimensional subspace of $\mathfrak{M}_{22}(K)$, but with the property that none of the matrices of B has an inverse. If we set $\mathfrak{S}_1 := \{\sigma(A) \mid \sigma \in GL(2,K)\}$ and $\mathfrak{S}_2 := \{\sigma(B) \mid \sigma \in GL(2,K)\}$, then $\mathfrak{S}_1 \cup \mathfrak{S}_2$ is a set of 2-dimensional vector subspaces such that the corresponding lines of the projective space form a spread $\mathfrak{S}$.

For the affine case the following two problems arise:

P7.3 Are there non-affine incidence spaces $(P,\mathfrak{L})$ with $\mathfrak{L}_{su} = \mathfrak{L}_1$, $k \geq 4$ and

a) $P_{II} \neq \emptyset$

b) $\mathfrak{L}_1$ is connected and contains a spread?

P7.4 In order to prove that, for $k \geq 4$ and $\mathfrak{L}_{su} = \mathfrak{L}_1$, $(P,\mathfrak{L})$ is an affine space, what assumptions do we have to make about the size of P_{II}?

P7.5 The same questions as in **P7.1**, **P7.2** and **P7.3** for the more general situation $\mathfrak{L}_{su} = \mathfrak{L}_i$ with $i > 1$.

8. (MDS)-Codes and Incidence Spaces with Projective and Affine Lines

In this section we will consider incidence spaces $(P,\mathfrak{L})$ of type $S(2,k)$ where $\mathfrak{L}_0$ resp. $\mathfrak{L}_1$ is not empty. We know that if $\mathfrak{L} = \mathfrak{L}_0$ (i.e. $(P,\mathfrak{L})$ is a projective space), then $v = |P|$ is a number of the form $v = 1 + n + n^2 + \ldots + n^d$ with $n := k-1$ and if $\mathfrak{L} = \mathfrak{L}_1$ then $v = k^d$. Further, if $(P,\mathfrak{L})$ is generated by 3 non-collinear points, then $\mathfrak{L}_0 \neq \emptyset$ implies that $(P,\mathfrak{L})$ is a projective plane, and $\mathfrak{L}_1 \neq \emptyset$ implies that

$(P,\mathfrak{L})$ is an affine plane. Therefore let us assume that $(P,\mathfrak{L})$ is not a plane.

If $\mathfrak{L}_0 \neq \emptyset$ then $(P,\mathfrak{L})$ contains small uniform books consisting of projective planes. Let $A \in \mathfrak{L}_0$. Then A is of type $[0,1]$ and $e_A \geq k$. If $e_A = k$ and if we put $n := k-1$, $v = k + e_A(|E|-k) = 1+n+(1+n)(1+n+n^2-(1+n)) = 1+n+n^2+n^3$ and this is the number of points of the 3-dimensional projective space of order n. For $e_A = k+1 = n+2$, $v = 1+n+(2+n)\cdot n^2 = 1+n+n^2+n^3+n^2$ is not any more the number of points of a projective space. Here we will deal with the tasks:

Construct incidence spaces $(P,\mathfrak{L})$ of type $S(2,n+1)$ with $\mathfrak{L}_0 \neq \emptyset$ such that

P8.1 $v = |P| = 1+n+n^2+n^3$ and $(P,\mathfrak{L})$ is not a projective space.

P8.2 $v = |P| = 1+n+n^2+n^3+n^2$.

For $\mathfrak{L}_1 \neq \emptyset$ and $A \in \mathfrak{L}_1$, $\mathfrak{C}_S(A)$ is a small uniform book of affine planes and A is of a type $[1,1]$. Here $e_A = k$ is the smallest possible value and then by **(7.2)b)**, $v = k(1+k(k-1)) = k^3-k^2+k$ which is not the number of points of an affine space of order k. We have again the problem:

P8.3 Is it possible to construct examples of incidence spaces of type $S(2,k,v)$ with $v = k^3-k^2+k$ and $\mathfrak{L}_1 \neq \emptyset$?

Another question is the following: Are there incidence spaces $(P,\mathfrak{L})$ with $\mathfrak{L}_0 \neq \emptyset$ and $\mathfrak{L}_1 \neq \emptyset$? In this case by **(7.2)** $v \equiv 0 \mod k$ and if $A \in \mathfrak{L}_0$, $B \in \mathfrak{L}_1$, $E \in \mathfrak{C}_S(A)$ and $F \in \mathfrak{C}_S(B)$ then $|E|-k = (k-1)^2$, $|F|-k = k(k-1)$ and so $v-k = e_A(|E|-k) = e_A(k-1)^2 = e_B(|F|-k) = e_B k(k-1)$, hence $e_A(k-1) = e_B\cdot k$, i.e. $e_A = \lambda k$ and $e_B = \lambda(k-1)$ with $\lambda \in \mathbb{N}$. Since $e_B \geq k$, $\lambda = 2$ and so $v = k+2k(k-1)^2 = 2k^3-4k^2+3k$ are the smallest possible values.

These problems can be translated into the language of codes. In order to do so, we need the notion of an (MDS)-**code** (**maximum distance sparable code**):
A q-nary code $\mathfrak{C} \subset K^m$ of length m over an alphabet K with $|K| = q$ letters is called an (MDS)- code if for any two **basic words** λe_i and μe_j with $\lambda,\mu \in K$, $i,j \in \{1,2,\dots,m\}$, $i \neq j$ there is exactly one code word $C \in \mathfrak{C}$ which is incident with λe_i and μe_j (i.e. the i-th position of C is occupied by the letter λ and the j-th position of C by the letter μ). The minimal distance of an (MDS)-code is $q-1$.

Let us first assume there is an incidence space $(P,\mathfrak{L})$ with a small uniform book $\mathfrak{C}_S(A)$ with $e_A = k$ pages as asked for in the problems **P8.1** and **P8.2**. Then every line $X \in \mathfrak{L}$ which is not contained in a page of the book $\mathfrak{C}_s(A)$ intersects each plane $E \in \mathfrak{C}_s(A)$ in a point of the set $E\backslash A$. Since for each $E \in \mathfrak{C}_s(A)$ the number $q := |E\backslash A| = |E|-k = (\gamma-1+\varepsilon k)(k-1)$ is constant we can identify each of these points $E\backslash A$ with an alphabet K consisting of q letters. Since $e_A = k$ we have k "generators" $\mathfrak{G} = \{E\backslash A \mid E \in \mathfrak{C}_s(A)\} =: \{E_1, E_2, \ldots, E_k\}$ and the set $P\backslash A = E_1 \dot\cup E_2 \dot\cup \ldots \dot\cup E_k$ can be identified with K^k. Then the set $\mathfrak{C} := \mathfrak{L}\backslash(\cup\{\mathfrak{L}(E) \mid E \in \mathfrak{C}_s(A)\})$ of all lines which are not contained in a plane E of the small uniform book $\mathfrak{C}_s(A)$ forms a q-nary (MDS)code of length k.

We have the result:

(8.1) Let $\mathfrak{C}_s(A)$ be a small uniform book with $k = e_A := |\mathfrak{C}_s(A)|$, let $\mathfrak{G} := \{E_1, \ldots, E_k\} := \{E\backslash A \mid E \in \mathfrak{C}_s(A)\}$, $q := |E_1|$ and $\mathfrak{C} := \{L \in \mathfrak{L} \mid \forall E \in \mathfrak{C}_s(A): |L \cap E| \le 1\}$; then:

a) $(P\backslash A, \mathfrak{G}, \mathfrak{C})$ is a 1-structure, i.e. a 1-chain structure with the property: For any two points $x,y \in P\backslash A$ with $[x] \ne [y]$ there is exactly one chain $C \in \mathfrak{C}$ such that $x,y \in C$, (i.e. $(P\backslash A, \mathfrak{G} \cup \mathfrak{C})$ forms an incidence space).

b) $\mathfrak{C}$ is a q-nary (MDS)-code of length k if we identify $P\backslash A$ with K where K is an alphabet with q letters.

These considerations show us, that to an incidence space $(P,\mathfrak{L})$ with a small uniform book $\mathfrak{C}_s(A)$ with $k = |\mathfrak{C}_s(A)|$ there correspond the following datas:

1. k planes $F_1, F_2, \ldots, F_k$ with the same number $|F_i| = 1 + (\gamma + \varepsilon k)(k-1)$ of points.

2. A q-nary (MDS)-code $\mathfrak{C}$ of length k with $q = (\gamma + \varepsilon k - 1)(k-1)$.

Therefore we assume vice versa that these two datas are given. In each plane F_i we distinguish a line A_i. Since all these lines A_i $(i \in \{1,2,\ldots,k\})$ have the same cardinality k we identify the points of all these lines and so obtain a single line A which is then the intersection of any two of the planes F_i, F_j with $i \ne j$. The set $P := F_1 \cup F_2 \cup \ldots \cup F_k$ has then $v = k + k(\gamma + \varepsilon k - 1)(k-1)$ points.

We set $E_i := F_i\backslash A$ and $\mathfrak{G} := \{E_1, E_2, \ldots, E_k\}$. Then $|E_i| = q$, $P\backslash A = E_1 \dot\cup E_2 \dot\cup \ldots \dot\cup E_k$, i.e. $(P\backslash A, \mathfrak{G})$ is a 1-web, and if K is the alphabet of the q-nary code $\mathfrak{C}$, there are k bijections

$$\pi_i : K \to E_i \quad \text{and} \quad \pi : \begin{cases} K^k & \to \binom{P\backslash A}{k} \\ (\lambda_1, \lambda_2, \ldots, \lambda_k) & \to \{\pi_1(\lambda_1), \pi_2(\lambda_2), \ldots, \pi_k(\lambda_k)\} \end{cases} \text{ is an injection}$$

of the all code K^n in the set of all k-sets of $P\setminus A$, characterized by $|\pi(\lambda_1,\lambda_2,\ldots,\lambda_k)\cap E_i| = 1$ for all $i \in \{1,2,\ldots,k\}$. Therefore for any code $\mathfrak{K} \subset K^k$, $(P\setminus A, \mathfrak{G}, \pi(\mathfrak{K}))$ is a 1-chain structure. For our given (MDS)-code $\mathfrak{C} \subset K^k$, $(P\setminus A, \mathfrak{G}\cup\pi(\mathfrak{C}))$ is an incidence space: For let $x,y \in P\setminus A$ be two distinct points. If $[x] = [y]$ then $[x] \in \mathfrak{G}$ is the only block which is incident with x and y; if $[x] \neq [y]$ there are $i,j \in \{1,2,\ldots,k\}$, $i \neq j$ with $[x] = E_i$ and $[y] = E_j$, and since $\mathfrak{C}$ is an (MDS)- code there is exactly one code word $C \in \mathfrak{C}$ which is incident with the basic words $\pi_i^{-1}(x)e_i$ and $\pi_j^{-1}(y)e_j$ and so $\pi(C)$ is the only block which contains x and y.

Finally we set $\mathfrak{L}^i := \{L \in \mathfrak{L}(F_i) \mid L \neq A\}$ and $\mathfrak{L} := \{A\} \cup \mathfrak{L}^1 \cup \mathfrak{L}^2 \cup \ldots \cup \mathfrak{L}^k \cup \pi(\mathfrak{C})$. Then $(P,\mathfrak{L})$ is an incidence space of type $S(2,k,k+kq)$ and $\mathfrak{E}_s(A) = \{F_1,\ldots,F_k\}$ is a small uniform book of $(P,\mathfrak{L})$: Let $x,y \in P$, $x \neq y$. If there is an $F_i \in \mathfrak{E}_s(A)$ with $x,y \in F_i$ then there is exactly one line $L \in \mathfrak{L}(F_i)$ with $x,y \in L$, and if there is not such plane F_i then $x,y \in P\setminus A$, $[x] \neq [y]$ and therefore there is exactly one $C \in \mathfrak{C}$ with $x,y \in \pi(C)$.

So we have:

(8.2) Let be given

1. k planes $F_1,\ldots,F_k$ with $|F_i| = 1+(\gamma+\varepsilon k)(k-1)$.

2. An (MDS)- code $\mathfrak{C}$ of length k over an alphabet K with $q := (\gamma+\varepsilon k-1)(k-1) = |K|$.

3. For each $i \in \{1,2,\ldots,k\}$ let $A_i \in \mathfrak{L}(F_i)$ and $\pi_i : K \to E_i := F_i\setminus A_i$ be a bijection. If we identify each line A_i with a fixed k-set A and if we set $P := A \cup E_1 \cup E_2 \cup \ldots \cup E_k$ and $\mathfrak{L} = \{A\} \cup \mathfrak{L}^1 \cup \mathfrak{L}^2 \cup \ldots \cup \mathfrak{L}^k \cup \pi(\mathfrak{C})$ where $\mathfrak{L}^i := \{L \in \mathfrak{L}(F_i) \mid L \neq A_i\}$ and $\pi : K^k \to \binom{P}{k}$; $(\lambda_1,\lambda_2,\ldots,\lambda_k) \to \{\pi_1(\lambda_1),\pi_2(\lambda_2),\ldots,\pi_k(\lambda_k)\}$, then $(P,\mathfrak{L})$ is an incidence space of type $S(2,k,k(q-1))$ and $\mathfrak{E}_s(A) = \{E_1 \cup A,\ldots,E_k \cup A\}$ a small uniform book.

Remarks.

1. In order to construct incidence spaces $(P,\mathfrak{L})$ of type $S(2,k,k((\gamma+\varepsilon k-1)(k-1)-1))$ which have a small uniform book $\mathfrak{E}_s(A)$ with $|\mathfrak{E}_s(A)| = k$, we have by **(8.1)** and **(8.2)** first to know that there are planes E with $|E| = 1+(\gamma+\varepsilon k)(k-1)$ points, and then we have to find an (MDS)-code $\mathfrak{C}$ of length k over an alphabet K with $|K| = = q := (\gamma - 1 + \varepsilon k)(k-1)$ letters. Then we are sure that such structures do exist.

2. For problem **P8.1** the planes E of the book $\mathfrak{C}_s(A)$ have to be projective. Therefore $\gamma = 0$ and $\varepsilon = 1$. Since we do not know whether there are projective planes of order $n := k-1$ if n is not the power of a prime, we will assume that n is a prime power. Then $q = (k-1)^2 = n^2$ is also a prime power, and we are allowed to consider the alphabet K as the finite field with q elements, and so we have to look for (MDS)-codes $\mathfrak{C}$ of the vector space (K^{n+1}, K).

3. In problem **P8.3** we will also assume that the order k of the affine planes is a prime power. Here $\gamma = \varepsilon = 1$ and so $q = k(k-1)$ is never the power of a prime hence the desired (MDS)-code $\mathfrak{C}$ can not be linear which makes our task much more difficult.

(MDS)-Codes and k-Webs

The problem to construct (MDS)-codes can also be translated into the language of k-webs: Let $\mathfrak{C} \subset K^k$ be an (MDS)-code of length k. Firstly we consider $\mathfrak{C}$ as the set of points, secondly for each $i \in \{1,2,\dots,k\}$ and for each $a \in K$ we call the set $G_{ia} := \{x_1, x_2, \dots, x_k\} \in \mathfrak{C} \mid x_i = a\}$ a generator and finally we set $\mathfrak{G}_i := \{G_{ia} \mid a \in K\}$. Then $(\mathfrak{C}, \mathfrak{G}_1 \cup \mathfrak{G}_2 \cup \dots \cup \mathfrak{G}_k)$ is a k-web and each generator G_{ia} contains exactly $q := |K|$ points of $\mathfrak{C}$. Therefore, if we are able to construct a k-web, $(P, \mathfrak{G}_1 \cup \mathfrak{G}_2 \cup \dots \cup \mathfrak{G}_k)$ with $k \geq 3$ and $|\mathfrak{G}_i| = q$ for $i \in \{1,2,\dots,k\}$ then $\mathfrak{C} := \left\{\left(\pi_1([p]_1), \pi_2([p]_2), \dots, \pi_k([p]_k)\right) \middle| p \in P\right\}$ is an (MDS)-code, if $\pi_i : \mathfrak{G}_i \to K$ is a bijection: for if $i, j \in \{1,2,\dots,k\}$, $i \neq j$ and $\lambda, \mu \in K$ are given then $\pi_i^{-1}(\lambda) \in \mathfrak{G}_i$ and $\pi_j^{-1}(\mu) \in \mathfrak{G}_j$ are two generators which intersect in exactly one point $p := \pi_i^{-1}(\lambda) \cap \pi_j^{-1}(\mu)$ since $i \neq j$, and $(\pi_1([p]_1), \pi_2([p]_2), \dots, \pi_k([p]_k))$ is the uniquely determined code word sought after.

(MDS)-Codes and Parallelisms

Let $\mathfrak{C} \subset K^k$ be an (MDS)-code and $q := |K|$. If $\lambda \in K$ and $i \in \{1,2,\dots,k\}$ then $|\mathfrak{C}(\lambda e_i)| = q$ for $\mathfrak{C}(\lambda e_i) := \{C \in \mathfrak{C} \mid \lambda e_i \in C\}$. If $D \in \mathfrak{C} \setminus \mathfrak{C}(\lambda e_i)$ and $(\lambda e_i \parallel D) := \{C \in \mathfrak{C}(\lambda e_i) \mid C \cap D = \emptyset\}$ then obviously $|(\lambda e_i \parallel D)| = q - (k-1)$ and so $q \geq k-1$. Let $(P, \mathfrak{G}, \mathfrak{C})$ be the corresponding chain structure and let $\mathfrak{L} := \mathfrak{G} \cup \mathfrak{C}$. Then $q = k-1$ if and only if $(P, \mathfrak{L})$ is a punctured projective plane; in this case $\mathfrak{L}_0 =$

$\mathfrak{C}$ and $\mathfrak{L}_1 = \mathfrak{G}$, and so we do not have a parallelism on $\mathfrak{C}$. If $q = k-2$ then $(\lambda \mathfrak{e}_i \parallel D)$ consists of exactly one line. Thus we have $q = k-2$ if and only if $(P,\mathfrak{L})$ is an affine plane, and then $\mathfrak{C}$ has the natural parallelism so that $\mathfrak{C}$ can be extended to an (MDS)-code of length $k+1$ which leads to a punctured projective plane.

By a **parallelism** $\parallel'$ on $\mathfrak{C}$ we understand an equivalence relation on $\mathfrak{C}$ such that the Euclidean parallelaxiom is true, i.e. if $C \in \mathfrak{C}$ and $p \in P$ then there is exactly one $D \in \mathfrak{C}$ with $p \in D$ and $D \parallel' C$. If $k > 3$ then our (MDS)-code $\mathfrak{C}$ can be reduced to an (MDS)-code $\mathfrak{C}'$ of length $k-1$ and $\mathfrak{C}'$ can be provided with a parallelism. For this purpose we have just to delete the points of one generator $E \in \mathfrak{G}$ and to consider two code-words $C \setminus E,\ D \setminus E \in \mathfrak{C}'$ with $C, D \in \mathfrak{C}$ as parallel if $C \cap D \cap E \neq \emptyset$. On the other hand if we have an (MDS)-code $\mathfrak{C}$ with parallelism then $\mathfrak{C}$ can be enlarged to an MDS)-code of length $k+1$. However, for $q \geq k-3$ it is very difficult to decide whether $\mathfrak{C}$ owns a parallelism, or what is the same, whether $\mathfrak{C}$ can be enlarged to an (MDS)code of length $k+1$.

Automorphisms of Incidence Spaces with a Small Book.

Let $(P,\mathfrak{L})$ be an incidence space of type $S(2,k)$ with $\mathfrak{L}_{su} \neq \emptyset$. It can happen that $\mathfrak{L}_{su}$ contains exactly one line A and then there is exactly one small uniform book $\mathfrak{C}_s(A)$. In any case if we select a line $A \in \mathfrak{L}_{su}$ then we can add to $(P,\mathfrak{L})$ an additional structure $\mathfrak{C}_s(A)$, i.e. we consider $(P,\mathfrak{L},\mathfrak{C}_s(A))$, and we can ask for the structure of the automorphism group $\Gamma := \mathrm{Aut}(P,\mathfrak{L},\mathfrak{C}_s(A))$. Γ consists of all collineations γ of $(P,\mathfrak{L})$ with $\gamma(A) = A$ and contains the subgroup $\Gamma' := \{\gamma \in \Gamma \mid \forall E \in \mathfrak{C}_s(A): \gamma(E) = E\}$. If $e_A := |\mathfrak{C}_s(A)| = k$ then, by theorem (**8.1**), we can consider the corresponding (MDS)- code $\mathfrak{C}$ and the restriction $\Gamma|_{P\setminus A}$ consists of automorphisms of the code $\mathfrak{C}$. The restriction $\Gamma|_{P\setminus A}$ is faithful and $\Gamma'|_{P\setminus A}$ are the code automorphisms which fix each generator of the code. Therefore Γ can be considered as a subgroup of the automorphism group $\mathrm{Aut}(\mathfrak{C})$ of the (MDS)-code $\mathfrak{C}$ and the question arises:

P8.4 When do we have $\Gamma = \mathrm{Aut}(\mathfrak{C})$?

The group Γ is contained in the bigger group
$\Sigma := \mathrm{Aut}(\mathfrak{C}_s(A)) := \{\sigma \in S_P \mid \forall X \in \mathfrak{L}_A: \sigma(X) \in \mathfrak{L}_A\}$ where $\mathfrak{L}_A := \cup \{\mathfrak{L}(E) \mid E \in \mathfrak{C}_s(A)\}$.
Of course, if $\sigma \in \Sigma$ then $\sigma(A) = A$ and $\sigma(E) \in \mathfrak{C}_s(A)$ for each $E \in \mathfrak{C}_s(A)$.

Each $\tau \in \mathrm{Aut}(\mathfrak{C})$ induces a permutation of the set $P\backslash A$ such that to each plane $E \in \mathfrak{C}_s(A)$ there is a plane $F \in \mathfrak{C}_s(A)$ with $\tau(E\backslash A) = F\backslash A$. Therefore we have $\tau \in \Gamma$ if and only if there is a $\sigma \in \Sigma$ such that $\sigma|_{P\backslash A} = \tau$. We can also consider the problem, given a subgroup $T \le \Sigma$, construct an (MDS)-code $\mathfrak{C}$ for $\mathfrak{C}_s(A)$ such that T is a subgroup of $\mathrm{Aut}\,\mathfrak{C}$ or even more of Γ. In the particular case that $A \in \mathfrak{L}_0$ or $A \in \mathfrak{L}_1$, i.e. $\mathfrak{C}_s(A)$ consists of projective or affine planes E, one would have to assume that the collineation groups of the planes E are not too small, for instance, one could demand that the planes are translation planes.

Problems and Solutions Concerning Parallelity

Given an affine plane E of order k. Then we can construct an incidence space $(P,\mathfrak{L})$ of type $S(2,k,\ k+(k^2-k)k)$ with $\mathfrak{L}_1 \neq \emptyset$ as soon as we can find an (MDS)-code $\mathfrak{C} \subset K^k$ of length k with $|K| = q = k^2-k$. If $k = 3$ then there are as many such (MDS)-codes as we can define loop operations on a set K of six elements (cf. § 10). But what about $k>3$?

P8.5 Construct (MDS)-codes $\mathfrak{C} \subset K^k$ with $q := |K| = k^2-k$ for $k>3$.

Let us now assume there is such an incidence code. Then there is still much freedom to construct incidence spaces of type $S(2,k,k(1+k^2-k))$ with $\mathfrak{L}_1 := \emptyset$. This leads to the problem:

P8.6 Is it possible to construct such an incidence space which admits a parallelism?

We discuss here the problem **P8.6** for $k = 3$. Then $q = 6$, $v = 21$ and the incidence structure of E is fixed because there is only one affine plane of order 3. We have given three affine planes E_i of order 3 which intersect in a line A. Let $A = \{a_1,a_2,a_3\}$ and $B_1 \in \mathfrak{L}(a_1,E_1)$, $B_1 \neq A$. First we assume there is a parallelism in $(P,\mathfrak{L})$. Each parallel class consists of 7 lines. If $L \in \mathfrak{C}$ then $[L]_{\parallel}$ contains at least one line X which intersects the spine A. If $X = A$ then $[L]_{\parallel}\backslash\{A\} \subset \mathfrak{C}$ and for any $M \in \mathfrak{C}\backslash[L]_{\parallel}$ ($\mathfrak{C}$ consists of 36 lines!) there are exactly three lines in $[M]_{\parallel}$ which intersect A, i.e. $[M]_{\parallel}$ consists of 4 lines of $\mathfrak{C}$. But $36-6 = 30$ is not divisible by 4. Therefore for each $L \in \mathfrak{C}$ $[L]_{\parallel}$ consists of exactly 4 lines of $\mathfrak{C}$ and 3 lines of $\mathfrak{L}_A$, and the parallels of A have to lie in the planes E_1,E_2,E_3.

Finally let $L \in \mathfrak{C}$ and let $B, C, D \in [L]_{\|}$ be the 3 lines with $B \cap A$, $C \cap A$, $D \cap A \neq \emptyset$. Suppose two of them are contained in one of the planes E_i, for instance $B, C \subset E_1$. But then $|E_1 \setminus (A \cup B \cup C)| = 9 - 7 = 2$ and this is a contradiction because the other 4 lines of $[L]_{\|}$ have to intersect $E_1 \setminus (A \cup B \cup C)$ in 4 distinct points. So we have the result:

If $(P, \mathfrak{L})$ is an incidence space of type $S(2,3,21)$ with $\mathfrak{L}_1 \neq \emptyset$ and a parallelism $\|$ then:

1. If $A \in \mathfrak{L}_1$ then $[A]_{\|}$ consists of A and the 6 lines of the affine book $\mathfrak{C}_s(A) = \{E_1, E_2, E_3\}$ which are contained in one of the affine planes E_i and which are parallel to A in E_i.

2. Let $A = \{a_1, a_2, a_3\}$. Each line $B_1 \in [a_1] \setminus \{A\}$ determines a parallel class consisting of exactly 4 lines of $\mathfrak{C}$ and three lines B_1, B_2, B_3 with $B_i \cap A = \{a_i\}$ and for $i \neq j$, B_i and B_j are not contained in one of the planes E_1, E_2, E_3.

In order to solve our problem one has firstly to consider decompositions of the code $\mathfrak{C}$ in 9 parallel classes, each class consisting of 4 code words. Secondly, if we denote the elements of K by $\{0, \pm 1, \pm 2, 3\}$ we must try to find a solution where in the rows of the blocks no 3 points are collinear.

The following table gives a decomposition of the code $\mathfrak{C} := \{(x, y, x+y) \mid x, y \in \mathbb{Z}_6\}$ in nine parallel classes where the second condition is fulfilled for the first row but not for the second.

$$
\begin{matrix} 0 & 0 & 0 \\ 2 & 1 & 3 \\ -1 & -1 & -2 \\ 3 & 2 & -1 \end{matrix}
\qquad
\begin{matrix} 0 & 2 & 2 \\ 2 & 3 & -1 \\ -1 & 1 & 0 \\ 3 & 0 & 3 \end{matrix}
\qquad
\begin{matrix} 0 & -1 & -1 \\ 2 & -2 & 0 \\ -1 & 2 & 1 \\ 3 & 1 & -2 \end{matrix}
\qquad
\begin{matrix} 0 & 1 & 1 \\ 1 & 3 & -2 \\ 2 & 0 & 2 \\ -2 & -1 & 3 \end{matrix}
\qquad
\begin{matrix} 0 & 3 & 3 \\ 1 & -2 & -1 \\ 2 & -1 & 1 \\ -2 & 2 & 0 \end{matrix}
$$

$$
\begin{matrix} 0 & -2 & -2 \\ 1 & 0 & 1 \\ -2 & 1 & -1 \\ 3 & -1 & 2 \end{matrix}
\qquad
\begin{matrix} 1 & 1 & 2 \\ 2 & 2 & -2 \\ -1 & -2 & 3 \\ -2 & 3 & -1 \end{matrix}
\qquad
\begin{matrix} 1 & -1 & 0 \\ -1 & 3 & 2 \\ -2 & 0 & -2 \\ 3 & -2 & 1 \end{matrix}
\qquad
\begin{matrix} 1 & 2 & 3 \\ -1 & 0 & -1 \\ -2 & -2 & 2 \\ 3 & 3 & 0 \end{matrix}
$$

P8.7 Given a projective plane E (desarguesian or not desarguesian) of order $n = k - 1$. If n is a prime power then there are (MSD)-codes $\mathfrak{C} \subset K^k$ with $q := |K| = (k-1)^2 = n^2$ (cf. Remark of (10.3)), and so by **(8.2)** there are incidence spaces $(P, \mathfrak{L})$ of type $S(2, 1+n, 1+n+n^2+n^3)$ with $\mathfrak{L}_0 \neq \emptyset$ such that all planes of

$\mathfrak{C}_s(A)$ are isomorphic to E. Is it possible to provide $(P,\mathfrak{L})$ with a parallelism?

Let $A \in \mathfrak{L}_0$ and let $\mathfrak{C}_s(A) = \{E_1,\dots,E_k\}$. We assume $(P,\mathfrak{L})$ admits a parallelism $\|$. Since E_i are projective planes, $[A]_\|$ consists of A and n^2 lines of $\mathfrak{C}$, i.e. we have a set of n^2 disjoint code words. Let $\mathfrak{C}' := \mathfrak{C}\setminus[A]_\|$; then $|\mathfrak{C}'| = n^2(n^2-1)$.

Let $B \in \mathfrak{L}_A$; then $[B]_\|$ has exactly $k = n+1$ lines which intersect A and $\frac{n^3-n}{n+1} = n(n-1)$ lines which are in $\mathfrak{C}'$. The $n(n-1)$ code words of $\mathfrak{C}' \cap [B]_\|$ are disjoint.

Therefore problem **P8.7** has the translation:

P8.7* Let $\mathfrak{C}$ be the (MDS)-code of $(P,\mathfrak{L})$
(hence $\mathfrak{C} \subset K^k$, $k = n+1$, $|K| = q = n^2 = (k-1)^2$, $|\mathfrak{C}| = q^2 = n^4$).

Decompose $\mathfrak{C} = \mathfrak{C}_r \,\dot\cup\, \dot{\bigcup\limits_{i=1}^{r-1}} \mathfrak{C}_i$ in $r = 1+n+n^2$ disjoint subsets such that

a) $\mathfrak{C}_r$ consists of $n^2 = (k-1)^2$ disjoint code words.

b) Foreach $i \in \{1,2,,\dots,r-1\}$, $\mathfrak{C}_i$ consists of $n^2-n = n(n-1) = (k-1)(k-2)$ disjoint code words.

Here we discuss this problem for $n = 2$, i.e. $k = 3$, $q = |K| = 4$ and $r = 7$. There are exactly the two loop operations:

1. $K = \{1,a,b,c\}$ is the Klein Four Group and the corresponding (MDS)- code $\mathfrak{C}$ has the decomposition ($|\mathfrak{C}_i| = n^2-n = 2$ for $i \in \{1,2,\dots,6\}$):

$$\mathfrak{C}_1\begin{cases}0\ a\ a\\ b\ 0\ b\end{cases}\qquad \mathfrak{C}_2\begin{cases}0\ b\ b\\ c\ 0\ c\end{cases}\qquad \mathfrak{C}_3\begin{cases}0\ c\ c\\ a\ 0\ a\end{cases}$$

$$\mathfrak{C}_4\begin{cases}a\ a\ 0\\ c\ b\ a\end{cases}\qquad \mathfrak{C}_5\begin{cases}b\ b\ 0\\ a\ c\ b\end{cases}$$

$$\mathfrak{C}_6\begin{cases}c\ c\ 0\\ b\ a\ c\end{cases}\qquad \mathfrak{C}_7\begin{cases}0\ 0\ 0\\ a\ b\ c\\ b\ c\ a\\ c\ a\ b\end{cases}$$

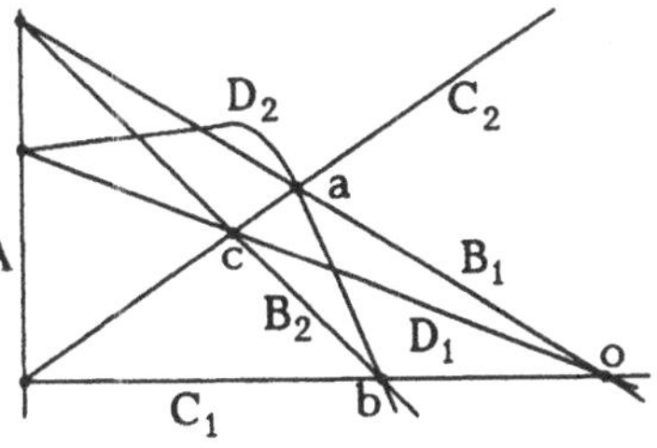

If we denote the lines of E as in the figure, and then the lines of $\mathfrak{L}_A$ in that way that the second index indicates the plane E_i, where the line is lying, we obtain the following parallel classes:

$[A]_{\|} = \{A\} \cup \mathfrak{C}_7$

$[B_{11}]_{\|} = \{B_{11}, C_{12}, D_{23}\} \cup \mathfrak{C}_6$, $[B_{21}]_{\|} = \{B_{21}, D_{22}, C_{13}\} \cup \mathfrak{C}_3$

$[B_{12}]_{\|} = \{B_{12}, D_{11}, C_{23}\} \cup \mathfrak{C}_5$, $[B_{22}]_{\|} = \{B_{22}, C_{21}, D_{13}\} \cup \mathfrak{C}_1$

$[B_{13}]_{\|} = \{B_{13}, D_{21}, C_{22}\} \cup \mathfrak{C}_2$, $[B_{23}]_{\|} = \{B_{23}, C_{11}, D_{12}\} \cup \mathfrak{C}_4$

2. $K = \mathbb{Z}_4$ is the cyclic group of order 4. Can we find a component $\mathfrak{C}_r = \mathfrak{C}_7$ with $|\mathfrak{C}_7| = 4$? If we assume $(0,0,0) \in \mathfrak{C}_7$ then there are exactly the 6 code words $(1,1,2)$, $(-1,-1,2)$, $(1,2,-1)$, $(2,1,-1)$, $(-1,2,1)$, $(2,-1,1)$ of $\mathfrak{C}$ which are disjoint to $(0,0,0)$. But as one checks, the maximal sets of disjoint code words containing $(0,0,0)$ have only 3 code words (there are six such sets). Since $(\mathbb{Z}_4,+)$ is commutative each code word $(a,b,a+b)$ defines by

$$(a,b,a+b)^+: \begin{cases} \mathfrak{C} \to \mathfrak{C} \\ (x,y,x+y) \to (a+x,\ b+y,\ a+b+x+y) \end{cases}$$

an automorphism of the code. All these automorphisms form a transitive group. If $(x,y,x+y) \parallel (u,v,u+v)$ then $(a+x,\ b+y,\ a+b+x+y) \parallel (a+u,\ b+v,\ a+b+u+v)$. This shows there is no set of 4 pairwise parallel code words. This gives us the result:

(8.3) All incidence spaces $(P,\mathfrak{L})$ of type $S(2,3,15)$ with $\mathfrak{L}_0 \neq \emptyset$ which result from an (MDS)-code $\mathfrak{C}$ derived from the Klein Four Group $\mathbb{Z}_2 \times \mathbb{Z}_2$ admit a parallelism, and those derived from the cyclic group $(\mathbb{Z}_4,+)$ do not admit a parallelism.

Also for the next problem we give a solution if $k=3$:

P.8.8 Find an incidence space $(P,\mathfrak{L})$ of type $S(2,k,k(k^2-k+1))$ with $|\mathfrak{L}_1| > 0$, $A,B \in \mathfrak{L}_1$, $A \neq B$ and $A \parallel B$; then $e_A = |\mathfrak{C}_s(A)| = |\mathfrak{C}_s(B)| = k$. $[A]_{\|} = [B]_{\|}$, $|[A]_{\|}| = k^2-k+1$ and $\forall E \in \mathfrak{C}_s(A)$, $\forall F \in \mathfrak{C}_s(B)$: $E = F$ or $E \cap F \in \mathfrak{L}$.
Solution for $k=3$: Let $\mathfrak{C}_s(A) = \{E_1,E_2,E_3\}$ and let $0_i \in E_i \setminus A$, $B_i := (0_i \parallel A)$. We denote $\{a_i,-a_i\} := B_i \setminus \{0_i\}$. In order to define a plane F_2 with $B_1,B_2,B_3 \subset F_2$ we obtain the loop:

0	a	-a
a	-a	0
-a	0	a

$B := B_1$ shall be a line of $\mathfrak{L}_1$. Let $b_i \in E_i \setminus (A \cup B_i)$ and $C_i := (b_i \parallel A)$.

Now $B_1 \cup C_2 \cup C_3$ shall be a plane F_3. Therefore we have to extend our loop: We need an element c such that $a + b = c$ where c_3 denotes a point of $C_3 \setminus \{b_3\}$ and an element d such that $-a + b = d$ and $C_3 = \{b_3, c_3, d_3\}$. Then C_2 must have the form $C_2 = \{b_2, c_2, d_2\}$ and we obtain so: $-a + d = c$, $a + c = d$, $-a + c = b$ and $a + d = b$.

0	a	-a	b	c	d
a	-a	0	c	d	b
-a	0	a	d	b	c

Now there are several possibilities to extend the partial operation to a loop operation on the set $K = \{0, a, -a, b, c, d\}$. We try to extend the multiplication in such a way, that also the line $C_1 = \{b_1, c_1, d_1\}$ becomes an affine line, i.e. $C_1 \in \mathfrak{L}_1$. Then C_1, B_2, C_3 and C_1, C_2, B_3 have to be complanar. The first condition implies $c + a = d$, hence $d - a = c$, $d + a = b$, $c - a = b$, $b + a = c$ and $b - a = d$ or $d + a = c$, hence $c - a = d$, $d - a = b$, $c + a = b$, $b + a = d$, $b - a = c$. This gives us the two extensions:

0	a	-a	b	c	d
a	-a	0	c	d	b
-a	0	a	d	b	c
b	c	d			
c	d	b			
d	b	c			

Tab. 1

0	a	-a	b	c	d
a	-a	0	c	d	b
-a	0	a	d	b	c
b	d	c			
c	b	d			
d	c	b			

Tab. 2

The second condition yields $\{-b, -c, -d\} \subset \{b, c, d\}$, for instance $-b = b$, $-c = c$, and then $-d = d$. Again we can put $b + c = a$ and then $b + d = d + c = -a$, $d + b = a$, $c + b = -a$, $c + d = a$ or $b + c = -a$ and then $b + d = d + c = a$, $c + d = d + b = -a$, $c + b = a$. So we can fill in in one of the empty spaces of Tab. 1 or Tab. 2 one of the two following small tables:

	b	c	d
b	0	a	-a
c	-a	0	a
d	a	-a	0

	b	c	d
b	0	-a	a
c	a	0	-a
d	-a	a	0

But there are also solutions such that K becomes a commutative loop: We set again $b + b := 0$ but $b + c := c + b := a$ and then $b + d = d + b = -a$, $c + c = -a$, $c + d = d + c = 0$, $d + d = a$.

So by using Tab. 1 and putting $d := -c$ we have also the solution:

0	a	-a	b	c	-c
a	-a	0	c	-c	b
-a	0	a	-c	b	c
b	c	-c	0	a	-a
c	-c	b	a	-a	0
-c	b	c	-a	0	a .

Tab. 3

It turns out even that by each of these solutions we obtain $\mathfrak{L}_1 = \{A, B_1, B_2, B_3, C_1, C_2, C_3\}$ hence seven affine lines which are pairwise parallel and seven affine planes $\{E_1, E_2, E_3, F_2, F_3, F_4, F_5\}$ where $F_4 := C_1 \cup B_2 \cup C_3$, $F_5 := C_1 \cup C_2 \cup B_3$. If there is a further mininmal plane D, D must be a projective plane. But we have $\mathfrak{L}_0 = \emptyset$ because $18 = = 21-3$ is not divisible by $4 = 7-3$. If we define the (MDS)-code $\mathfrak{C}$ by the loop given by Tab. 3, then we see that the points $0_1, a_2, a_3$ resp. $0_1, c_2, c_3$ resp. b_1, c_2, a_3 resp. b_1, a_2, c_3 are collinear because $(0,a,a)$, $(0,c,c)$, (b,c,a) and (b,a,c) are code words. Since the incidence structure of $(P,\mathfrak{L})$ is not yet completely determined and since the lines $\overline{0_1, b_1}$, $\overline{a_2, c_2}$, $\overline{a_3, c_3}$ must lie in the planes E_1, E_2, E_3 resp., we may arrange that these lines intersect the line A in a common point u. Then $\{u, 0_1, b_1, a_2, c_2, a_3, c_3\}$ is a projective plane.

For an incidence space $(P,\mathfrak{L})$ of type $S(2,3)$ with $\mathfrak{L}_0 \neq \emptyset$ and $\mathfrak{L}_1 \neq \emptyset$ we obtain: If $A \in \mathfrak{L}_0$ and $B \in \mathfrak{L}_1$, then $v-3 = e_A \cdot 4 = e_B \cdot 6$, hence $v-3$ is divisible by 12, A and B are skew and $e_A, e_B \geq 3$. Therefore $v = 3+24 = 27$ is the smallest possible number of points of such an incidence space and then $e_A = 6$ and $e_B = 4$.

9. m-Incidence Codes

In problem **P.8.2** one is looking for incidence spaces $(P,\mathfrak{L})$ with a small uniform book $\mathfrak{E}_S(A)$ where the number $e_A := |\mathfrak{E}_S(A)|$ of pages is larger than k. This leads us to the notion of an m-incidence code, a generalization of the (MDS)-codes. If again $q := |E \setminus A|$ for $E \in \mathfrak{E}_S(A)$, let K be a set with $|K| = q$, and if $\mathfrak{E}_S(A) = \{E_1, E_2, \dots, E_e\}$, let $\pi_i : E_i \setminus A \to K$ be a bijection for each $i \in \{1, 2, \dots, e_A\}$.

This set K we "compactify" by adding an improper element ∞, hence $\overline{K} := K \cup \{\infty\}$. $\overline{K}$ will be our alphabet, where we consider ∞ as a distinguished element.

The length of our code will be the number e_A and we set $m := e_A - k$. Let $\mathfrak{C}$ be the set of all lines which are not contained in one of the planes $E_1, E_2, \dots, E_{e_A}$; then for

$L \in \mathfrak{C}$, $|L \cap (E_i \setminus A)| \le 1$ and we define $\pi_i(L) := \begin{cases} \pi_i(L \cap E_i) & \text{if } L \cap E_i \neq \emptyset \\ \infty & \text{if } L \cap E_i = \emptyset. \end{cases}$

Then $(\pi_1(L), \pi_2(L), \dots, \pi_{e_A}(L)) \in \overline{K}^{e_A}$ and exactly m of the positions are occupied by ∞. Hence $\mathfrak{C}' := \{(\pi_1(L), \pi_2(L), \dots, \pi_{e_A}(L)) \mid L \in \mathfrak{C}\}$ is a code of length e_A with the alphabet $\overline{K}$. Further, if $\lambda, \mu \in K$ and $i, j \in \{1, 2, \dots, e_A\}$ with $i \neq j$ then $\pi_i^{-1}(\lambda) \in E_i \setminus A$ and $\pi_j^{-1}(\mu) \in E_j \setminus A$, so that $L := \overline{\pi_i^{-1}(\lambda), \pi_j^{-1}(\mu)}$ is a line which belongs to $\mathfrak{C}$. Therefore $\mathfrak{C}'$ has exactly one code word C where the i-th position is occupied by λ and the j-th position by μ.

Definition. Let $\overline{K} = K \dot{\cup} \{\infty\}$, $k \in \mathbb{N}$, $k \ge 3$ and $m \in \mathbb{N}_0$. A code $\mathfrak{C} \subset \overline{K}^{(k+m)}$ is called an **m-incidence code** if the two conditions are valid:

1. If $C \in \mathfrak{C}$ then exactly m positions of the code word C are occupied by ∞.

2. If $i, j \in \{1, 2, \dots, k+m\}$, $i \neq j$ and $\lambda, \mu \in K$ then there is exactly one code word $C \in \mathfrak{C}$ such that the i-th position is occupied by λ and the j-th by μ.

Example: Find an incidence space $(P, \mathfrak{L})$ of type $S(2,3,19)$ with $\mathfrak{L}_0 \neq \emptyset$ and $e_A := |\mathfrak{C}_S(A)| = 4$ for $A \in \mathfrak{L}_0$. Here each plane $E \in \mathfrak{C}_S(A)$ has 7 points and so $|K| = 4$, $|\overline{K}| = 5$. A solution is here the following: Let $K = \{0, a, b, c\} = \mathbb{Z}_2 \times \mathbb{Z}_2$ be the Klein-Four-Group. Then we can extend the (MDS)-code corresponding to $\mathbb{Z}_2 \times \mathbb{Z}_2$ as follows (we put in the element ∞ and apply the translation b^+):

0	0	0	∞	0	b	∞	b	0	∞	b	0	∞	0	b	b
0	a	a	∞	0	c	∞	c	0	∞	c	a	∞	0	c	c
a	0	a	∞	a	b	∞	c	a	∞	b	a	∞	a	b	c
a	a	0	∞	a	c	∞	b	a	∞	c	0	∞	a	c	b
b	b	b	∞	b	0	∞	0	b	∞	0	b	∞	b	0	0
b	c	c	∞	b	a	∞	a	b	∞	a	c	∞	b	a	a
c	b	c	∞	c	0	∞	a	c	∞	0	c	∞	c	0	a
c	c	b	∞	c	a	∞	0	c	∞	a	b	∞	c	a	0

10. Construction of (MDS)-Codes

First we assume that $\mathfrak{C}$ is an (MDS)-code of length $k \geq 3$ over an alphabet $K = \{0, 1, \alpha_3, \ldots, \alpha_q\}$ with $q = |K|$ letters. We consider $\mathfrak{C}$ as the chains of a 1-chain structure $(P, \mathfrak{G}, \mathfrak{C})$ with $\mathfrak{G} = \{Ke_i \mid i \in \{1,2,\ldots,k\}\}$ and $P := Ke_1 \dot{\cup} Ke_2 \dot{\cup} \ldots \dot{\cup} Ke_k$ (cf. §4). Since $\mathfrak{C}$ is an (MDS)-code, $(P, \mathfrak{G}, \mathfrak{C})$ is even a 1-structure and so by (3.5), $k \leq q+1$. For $i,j \in \{1,2,\ldots,k\}$, $i \neq j$ and $\lambda, \mu \in K$ let $\overline{\lambda e_i, \mu e_j} = C$ be the uniquely determined chain $C \in \mathfrak{C}$ with $\lambda e_i, \mu e_j \in C$. To any triple (h,i,j) with $h,i,j \in \{1,2,\ldots,k\}$ and $i \neq h,j$ and any letter $\alpha \in K$ there corresponds exactly one permutation $\alpha_{hij}\colon K \to K$ defined by $\alpha_{hij}(\xi) = \eta$ if $\eta e_h = \overline{\xi e_j, \alpha e_i} \cap Ke_h$. We have $\alpha_{hij}^{-1} = \alpha_{jih}$ and $\alpha_{hij} \circ \alpha_{jim} = \alpha_{him}$ and therefore for each $\alpha \in K$ and each $i \in \{3,4,\ldots,k\}$, $\alpha_i^+ := 0_{12i} \circ \alpha_{i21}$ is a permutation of K and 0_i^+ is the identity.

We set $\Pi_i := \{\alpha_i^+ \mid \alpha \in K\}$. Then (Π_i, K) is a set of fixed point free permutations, (i.e. if $\alpha_i^+(\xi) = \xi$ for one $\xi \in K$ then α_i^+ is the identity) which acts regularly on K: For let $\xi, \eta \in K$ be given, then α is uniquely determined by $\alpha e_2 = \overline{\xi e_1, \eta' e_i} \cap Ke_2$ where $\eta' e_i = \overline{\eta e_1, 0 e_2} \cap Ke_i$.

Also if for $i,j \in \{3,4,\ldots,k\}$ and $i \neq j$, we define $\Pi_{ij} := \{\alpha_i^+ \circ (\alpha_j^+)^{-1} \mid \alpha \in K\}$ then the pair (Π_{ij}, K) is a regular permutation set: For let $\xi, \eta \in K$ be given. We set $\xi' := 0_{j21}(\xi)$, $\eta' := 0_{i21}(\eta)$ and define α by $\alpha e_2 = Ke_2 \cap \overline{\xi' e_j, \eta' e_i}$. Then $\alpha_{i2j}(\xi') = \eta'$ and $0_{12i}(\eta') = \eta$, hence $\alpha_i^+ \circ (\alpha_j^+)^{-1}(\xi) = 0_{12i} \circ \alpha_{i21} \circ \alpha_{12j} \circ 0_{j21}(\xi) = 0_{12i} \circ \alpha_{i2j}(\xi') = 0_{12i}(\eta') = \eta$.

We have the result:

(10.1) To each (MDS)-code $\mathfrak{C} \subset K^k$ there correspond $k-2$ regular permutation sets (Π_i, K) and $k-2$ bijections $\beta_i\colon K \to \Pi_i;\ \alpha \to \alpha_i^+$ (for $i \in \{3,4,\ldots,k\}$) such that

1. $\forall\, i \in \{3,4,\ldots,k\}$: $0_i^+ = \mathrm{id}$

2. $\forall\, i,j \in \{3,4,\ldots,k\}$, $i \neq j$: (Π_{ij}, K) with $\Pi_{ij} := \{\alpha_i^+ \circ (\alpha_j^+)^{-1} \mid \alpha \in K\}$ is a regular permutation set.

Remark. For $i,j \in \{3,4,\ldots,k\}$ $(K, \oplus_i)$ and if $i \neq j$, $(K, \oplus_{ij})$ are loops with respect to the operations

$$\oplus_i : \begin{cases} K \times K & \to K \\ (\alpha,\beta) & \to \alpha_i^+(\beta) \end{cases} \quad \text{and} \quad \oplus_{ij} : \begin{cases} K \times K & \to K \\ (\alpha,\beta) & \to \alpha_i^+ \circ (\alpha_j^+)^{-1}(\beta) \end{cases}.$$

Also the converse of (**10.1**) is true:

(**10.2**) Let be given:

a) A finite set K with $q := |K| \geq 2$ where an element $0 \in K$ is distinguished.

b) $k-2$ regular permutation sets (Π_i, K), $i \in \{3,4,\dots,k\}$ where $k \geq 3$.

c) $k-2$ bijections $\beta_i: K \to \Pi_i;\ \alpha \to \alpha_i^+$.

If the conditions **1.** and **2.** of (**10.1**) are valid then $\mathfrak{C} := \{(\alpha,\beta,\alpha_3^+(\beta),\dots,\alpha_k^+(\beta)| \alpha,\beta \in K\}$ is a q-nary (MDS)-code of length k with $\mathfrak{C} \subset K^k$ and $|\mathfrak{C}| = q^2$.

Proof. For $\alpha,\beta \in K$ let $\overline{(\alpha,\beta)} := (\alpha,\beta,\alpha_3^+(\beta),\dots,\alpha_k^+(\beta))$. Let $i,j \in \{1,2,\dots,k\}$ with $i \neq j$ and $\lambda,\mu \in K$ be given. If $i,j \neq 1,2$ then by condition **2.** there is exactly one $\alpha \in K$ such that $\alpha_i^+ \circ (\alpha_j^+)^{-1}(\mu) = \lambda$ so that for $\beta := (\alpha_j^+)^{-1}(\mu)$, $\overline{(\alpha,\beta)}$ is the only code word of $\mathfrak{C}$ which is incident with λe_i and μe_j. If $j \neq 2$ then by **b**) and **c**) there is exactly one $\alpha \in K$ such that $\alpha_j^+(\lambda) = \mu$, and $\overline{(\alpha,\lambda)}$ is the only element $C \in \mathfrak{C}$ with $\lambda e_2, \mu e_j \in C$; there is also exactly one element $\beta \in K$ with $\lambda_j^+(\beta) = \mu$ since λ_j^+ is a permutation, and so $\lambda e_1, \mu e_j \in \overline{(\lambda,\beta)}$. Finally $\lambda e_1, \mu e_2 \in \overline{(\lambda,\mu)}$.

Examples. For $k = 3$ the conditions of theorem (**10.2**) reduce to the demand that there is only one regular permutation set (Π, K) with $id \in \Pi$ or what is the same, that K can be provided with a loop operation "+". But since $q = |K|$ is finite there are even group operations, in particular, K can be considered as the cyclic group $(\mathbb{Z}_q,+)$. If $(K,\oplus)$ is a loop then $\mathfrak{C} := \{(x,y,x\oplus y)| \ x,y \in K\}$ is an (MDS)-code. In the case of problem **P8.1** we have $n = k-1 = 2$ and $q = 4$, and on a set of four elements, there are only two loop operations, the cyclic group $\mathbb{Z}_4$ and the Klein Four Group $\mathbb{Z}_2 \oplus \mathbb{Z}_2$.

Also there is only one projective plane E of order 2: $E = \mathbb{Z}_2^3 \setminus \{(0,0,0)\}$ and $\mathfrak{L}(E) = \{\{a,b,a+b\}| \ a,b \in E,\ a \neq b\}$. If we choose $A := \{(0,1,0),(0,0,1),(0,1,1)\}$, hence $E \setminus A = \{(1,0,0),(1,1,0),(1,0,1),(1,1,1)\}$ and three bijections $\pi_i: K \to E\setminus A$, then the corresponding incidence space $(P,\mathfrak{L})$, according to theorem (**8.2**), is a projective space if and only if the following conditions are true:

(*) $\quad \forall a,b,c \in K,\quad \pi_1(a) + \pi_1(a \oplus b \ominus c) = \pi_2(b) + \pi_2(c) = \pi_3(a \oplus b) + \pi_3(a \oplus c)$

(**) $(K,\oplus)$ is the Klein Four Group $\mathbb{Z}_2 \oplus \mathbb{Z}_2$.

Proof. If $b \neq c$ then $F := \overline{\{\pi_1(a), \pi_2(b), \pi_3(c)\}}$ is a plane which contains the points $\pi_3(a \oplus b), \pi_3(a \oplus c)$ and $B := \{\pi_2(b), \pi_2(c), \pi_2(b)+\pi_2(c)\}$ and $C := \{\pi_3(a\oplus b), \pi_3(a\oplus c), \pi_3(a\oplus b)+\pi_3(a\oplus c)\}$ are two lines with $B \subset F \cap E_2$, $C \subset F \cap E_3$ and $B \cap A = \{\pi_2(b)+\pi_2(c)\}$, $C \cap A = \{\pi_3(a\oplus b)+\pi_3(a\oplus c)\}$. If F is a minimal plane then $F \cap E_2 = B$, $F \cap E_3 = C$ and so $F \cap E_2 \cap E_3 = F \cap A = B \cap A = C \cap A$, hence $\pi_2(b)+\pi_2(c) = \pi_3(a\oplus b)+\pi_3(a\oplus c)$. If $b = c$ then $\pi_2(b)+\pi_2(c) = (0,0,0) = \pi_3(a\oplus b)+\pi_3(a\oplus c)$. Therefore, if $(P,\mathfrak{L})$ is a projective space then (*) is valid.

Let again $b \neq c$ and $d \in K$ such that $d\oplus b = a\oplus c$. Then $d \neq a$ and $\pi_1(d) \in \overline{\pi_2(b), \pi_3(a\oplus c)} \subset F$, hence $\pi_3(d\oplus c) \in \overline{\pi_1(d), \pi_2(c)} \subset F$ and $\pi_3(d\oplus c) \neq \pi_3(a\oplus c)$. If F is a projective plane then $\pi_3(d\oplus c) \in C\setminus A$ and, since $|C| = 3$ thus $C\setminus A = \{\pi_3(a\oplus b), \pi_3(a\oplus c)\}$, we have $\pi_3(d\oplus c) = \pi_3(a\oplus b)$, consequently $d\oplus c = a\oplus b$. Since $(K,\oplus)$ is a commutative group we obtain $a \oplus b \ominus c = d = a\oplus c \ominus b$, hence $b \oplus b = c \oplus c$. If $(P,\mathfrak{L})$ is a projective space, this equation holds for all $b,c \in K$ with $b \neq c$, hence also for $c = 0$ implying $b \oplus b = 0$ for all $b \in K$. Thus $(K,\oplus)$ is the Klein-Four-Group.

Now let us assume (*) and (**). We have to consider three non collinear points $p_1, p_2, p_3 \in P$ which are not contained in one of the three planes E_1, E_2, E_3 and the plane $F := \overline{\{p_1, p_2, p_3\}}$. If $L_1 := \overline{p_2, p_3}$, $L_2 := \overline{p_3, p_1}$, $L_3 := \overline{p_1, p_2}$ then $L_1, L_2, L_3 \neq A$ and therefore we may assume $p_1, p_2, p_3 \notin A$. At most one of the lines L_i can be contained in one of the planes E_1, E_2, E_3. So let us assume $L_1, L_2 \not\subset E_1, E_2, E_3$. Then there are $a,b,c \in K$ such that $L_1 \cap E_1 = \pi_1(a)$, $L_1 \cap E_2 = \pi_2(b)$, $L_1 \cap E_3 = \pi_3(a\oplus b)$ and $L_2 \cap E_2 = \pi_2(c)$, and consequently $F = \overline{\{\pi_1(a), \pi_2(b), \pi_3(c)\}}$. F contains the four further points $\pi_3(a\oplus b)$, $\pi_3(a\oplus c)$, $\pi_2(b)+\pi_2(c) = \pi_3(a\oplus b)+\pi_3(a\oplus c)$ and $\pi_1(a\oplus b\ominus c)$. If we put $B_1 := \overline{\pi_1(a), \pi(a\oplus b\ominus c)}$, $B_2 := \overline{\pi_2(b), \pi_2(c)}$, $B_3 := \overline{\pi_3(a\oplus b), \pi_3(a\oplus c)}$ then by (*) $B_1 \cap B_2 = B_2 \cap B_3 = \{\pi_2(b)+\pi_2(c)\}$ and for $C := \overline{\pi_1(a\oplus b\ominus c), \pi_2(b)}$ we have $C \cap B_3 = \{\pi_3(a\oplus c)\}$ since by (**) $a\oplus b \ominus c \oplus b = a \oplus c$. This shows that $(F,\mathfrak{L}(F))$ is a projective plane, and so $(P,\mathfrak{L})$ is a projective space. Since it is easy to violate one of the conditions (*) or (**) there are plenty of incidence spaces $(P,\mathfrak{L})$ with $|P| = 27$ and $\mathfrak{L}_0 \neq \emptyset$ which are not projective spaces.

Also problem **P8.3** can be easily solved for $k = 3$. Here $q = k^2 - k = 6$ and if we take $K := (\mathbb{Z}_6, +)$ then $\mathfrak{C} = \mathbb{Z}_6 \cdot (1,0,1) + \mathbb{Z}_6 \cdot (0,1,1)$ is an (MDS)-code.

For $k > 3$ the task, to construct (MDS)-codes becomes much more difficult. In [16] the problem will be discussed how one can find (MDS)-codes with the help of finite near-rings. A **near-ring** $(N,+,\cdot)$ is defined by (a comprehensive presentation of the theory of near-rings and near-fields is given in [17], [18] resp. [19]):

N.1 $(N,+)$ is a group; let 0 denote the neutral element

N.2 $(N,\cdot)$ is a semigroup

N.3 $\forall a,b,c \in N: \ (a+b)\cdot c = ac + bc$

and a near-ring $(N,+,\cdot)$ is a **near-field** if for $N^* := N \setminus \{0\}$, $(N^*,\cdot)$ is a group. If $(K,+,\cdot)$ is a finite field or more general a finite near-field, i.e. $q = |K|$ is the power of a prime, then we can give (MDS)-codes where k attains the maximal value $q+1$:

Let $K = \{\alpha_2 = 0, \ \alpha_3 = 1, \ \alpha_4, \dots, \alpha_{q+1}\}$, and for $\alpha \in K$ and $i \in \{3,4,\dots,k = q+1\}$ let $\alpha_i^+ : K \to K; \ \xi \to \alpha_i \cdot \alpha + \xi$. Then the conditions of (10.2) are fulfilled.

Hence we have the result:

(10.3) Let $(K,+,\cdot)$ be a finite near-field with $q = |K|$ and let $K = \{\alpha_1 = 0, \ \alpha_2 = 1, \ \alpha_3, \dots, \alpha_q\}$. Then $\mathfrak{C} := \{(\xi, \eta, \xi+\eta, \alpha_3 \cdot \xi + \eta, \dots, \alpha_q \xi + \eta) \mid \xi, \eta \in K\}$ is a q-nary (MDS)-code of maximal length $k = q+1$.

Remark. If q is the power of a prime then there is always a finite field K of order q. Therefore by (10.3) for each $k \in \mathbb{N}$ with $3 \leq k \leq q+1$ there are q-nary (MDS)-codes of length k.

11. Möbius Structures

In this section let $(P,\mathfrak{K})$ be a Möbius structure where $k^{(1)} = k \geq 3$ is constant. The blocks $X \in \mathfrak{K}$ we call here **circles** and $n := k-1$ the **order** of $(P,\mathfrak{K})$. Since by definition (cf. §3) $r^{(3)} = 1$ is constant, $(P,\mathfrak{K})$ is of type $S(3,k)$. So by §2 the derivation $(P^{(a)},\mathfrak{K}^{(a)})$ in a point $a \in P$ is an incidence space of type $S(2,k-1)$

and by (**2.4**), $r^{(2)} = \frac{v-2}{k-2} = \frac{v-2}{n-1}$, $r = r^{(2)} \cdot \frac{v-1}{k-1} = \frac{v-2}{n-1} \cdot \frac{v-1}{n}$ and $b = \frac{v-2}{n-1} \cdot \frac{v-1}{n} \cdot \frac{v}{n+1}$ are integers, so far $v := |P|$ is finite. We have for $(P, \mathfrak{K})$ the following parameter equations cf. (**2.4**), (**2.5**):

(**11.1**) **a**) $v-2 = r^{(2)}(k-2) = r^{(2)}(n-1)$, $r^{(2)}(r^{(2)}-1) \equiv 0 \mod n$ and $v \cdot r \equiv 0 \mod (n+1)$.

b) If the order $n = k-1$ is a prime power then there is an $\alpha \in \mathbb{N}$ such that either $r^{(2)} = \alpha \cdot n$ hence $r = \alpha(v-1)$, $v = 2 + \alpha n(n-1)$ and $2\alpha(1+\alpha)(1+2\alpha) \equiv 0 \mod (n+1)$ or $r^{(2)} = 1 + \alpha n$ hence $v = 2 + (1+\alpha n)(n-1)$, $r = (1+\alpha n)(1+\alpha n - \alpha)$ and $2\alpha(1-\alpha)(1-2\alpha) \equiv 0 \mod (n+1)$.

To the Möbius structures $(P, \mathfrak{K})$ of type $S(3,k)$ there belong the finite **Möbius-planes** which are characterized by $|\mathfrak{K}| > 1$ and the touching axiom **T** (cf. § 3) or by the condition that the derivation in each point $a \in P$ is an affine plane. Each Möbius plane of order n is of type $S(3, n+1, n^2+1)$ (cf. (**3.4**)) and has the further parameters $r^{(2)} = n+1 = k$, $r = (n+1) \cdot n = n^2 + n$ and $b = n(n^2+1)$. Vice versa every Möbius structure $(P, \mathfrak{K})$ of type $S(3, n+1, n^2+1)$ is a Möbius plane because the derivation $(P^{(p)}, \mathfrak{K}^{(p)})$ in a point $p \in P$ is an incidence space of type $S(2, n, n^2)$, hence by (**2.5**)**e**) an affine plane.

If the derivation of $(P, \mathfrak{K})$ in a point p is a projective plane, i.e. $r^{(2)} = k-1 = n$, then by (**11.1**) $v = 2 + n(n-1) \equiv 4 \mod (n+1)$, $r = v-1 \equiv 3 \mod (n+1)$, hence $v \cdot r \equiv 12 \equiv 0 \mod (n+1)$. Therefore n has to be one of the values 1, 2, 3, 5, 11. Since $n-1$ is the order of the projective plane, only the values 3, 5 remain and the value 11 if there is a projective plane of order 10. Thus $(P, \mathfrak{K})$ has to be of one of the types $S(3,4,8)$, $S(3,6,22)$ or $S(3,12,112)$. Möbius structures of type $S(3,4,8)$ do exist and they are all isomorphic: Let $P := (\mathbb{Z}_2)^3$ be the set of points and $\mathfrak{K}$ the set' of all planes of the 3-dimensional affine space $A((\mathbb{Z}_2)^3, \mathbb{Z}_2)$. Also there are Möbius structures of type $S(3,6,22)$ as E. WITT [20] has shown.

Thus we have the result:

(**11.2**) For a finite incidence structure $(P, \mathfrak{K})$ the following statements **a**), **b**), **c**), **d**) resp. **a'**), **b'**), **c'**) are equivalent.

a) $(P, \mathfrak{K})$ is a Möbius plane.

b) $(P,\mathfrak{K})$ is of type $S(3,k)$ and there is a $p \in P$ such that the derivation $\left(P^{(p)},\mathfrak{K}^{(p)}\right)$ is an affine plane.

c) $\forall p \in P$: $\left(P^{(p)},\mathfrak{K}^{(p)}\right)$ is an affine plane.

d) $(P,\mathfrak{K})$ is of type $S(3,k,k^2+1)$.

a') $(P,\mathfrak{K})$ is of type $S(3,k)$ and $\exists p \in P$ such that $\left(P^{(p)},\mathfrak{K}^{(p)}\right)$ is a projective plane.

b') $\forall p \in P$: $\left(P^{(p)},\mathfrak{K}^{(p)}\right)$ is a projective plane.

c') $(P,\mathfrak{K})$ is of one of the types $S(3,4,8)$, $S(3,6,22)$ or $S(3,12,112)$.

In a Möbius structure $(P,\mathfrak{K})$ of type $S(3,k)$ with $|X| \geq 4$ for all $X \in \mathfrak{K}$ we call a subset $M \subset P$, a **circular subspace** if with any three distinct points $a,b,c \in M$ also the circle X, determined by a,b,c, is contained in M. The set $\mathfrak{M}$ of all circular subspaces of $(P,\mathfrak{K})$ is $\cap$-closed, and so we have a circular closure operation. W.HEISE [21] calls a Möbius structure $(P,\mathfrak{K})$ a **Möbius-space** if for any four non concylic points their circular closure is a Möbius plane. There are infinite Möbius spaces of any dimension, but the only non trivial finite Möbius spaces are the Möbius planes.

The Möbius structures $(P,\mathfrak{K})$ where the derivation in a point is an affine space, have (in the finite case) the parameters $v = n^m + 1$, $r^{(2)} = 1 + n + n^2 + \ldots + n^{m-1}$, hence $\alpha = 1 + n + \ldots + n^{m-2}$. Now let us assume that the automorphism group of a Möbius structure $(P,\mathfrak{K})$ contains a subgroup Γ which acts sharply 3-transitive on the set P of points. For $\gamma \in \Gamma$ let $\mathrm{Fix}(\gamma) := \{x \in P \mid \gamma(x) = x\}$ and for $a \in P$ let $\Gamma_a := \{\gamma \in \Gamma \mid \gamma(a) = a\}$. Then we know (cf. e.g. [22]): If P is finite and if we fix three distinct points $0,1,\infty \in P$ then for each $a \in L := P\setminus\{\infty\}$ there is exactly one $a^+ \in \Gamma_\infty$ such that $a^+(0) = a$ and $\mathrm{Fix}(a^+) = \{\infty\}$ if $a \neq 0$, and exactly one $a^\cdot \in \Gamma_{(\infty,0)}$ such that $a^\cdot(1) = a$ and $\mathrm{Fix}(a^\cdot) = \{\infty,0\}$ if $a \neq 1$.

We put $0^+ := 1^\cdot := \mathrm{id}$. With respect to the operations

$$+: \begin{cases} L \times L \to L \\ (a,b) \to a^+(b) \end{cases} \qquad \cdot: \begin{cases} L \times L \to L \\ (a,b) \to \begin{cases} a^\cdot(b) & \text{if } a \neq 0 \\ 0 & \text{if } a = 0 \end{cases} \end{cases}.$$

$(L,+,\cdot)$ is a left **near-field** (i.e. $(L,+)$ is a commutative group, $(L^*,\cdot)$ for $L^* := L\setminus\{0\}$ is a group and $a(b+c) = ab + ac$ for all $a,b,c \in L$). If $\overline{K}$ denotes the circle through $0,\infty,1$ and $K := \overline{K}\setminus\{\infty\}$ then $(K,+,\cdot)$ is a subnear-field of

$(L,+,\cdot)$. The map $\sigma \in \Gamma_1$ determined by $\sigma(0) = \infty$ and $\sigma(\infty) = 0$ is an involution with $\sigma(\overline{K}) = \overline{K}$ and $\tau := \sigma|L^*$ is an isomorphism of the multiplicative group $(L^*,\cdot)$ which satisfies the functional equation

$$(*) \quad \tau(1+\tau(x)) = 1-\tau(1+x) \quad \text{for all } x \in L^*\setminus\{-1\}.$$

Such structures $(L,+,\cdot,\tau)$ are called **KT-near-fields** (cf. [22]). Since $\sigma(\overline{K}) = \overline{K}$ we can form the restriction $\tau' := \tau|K^*$, and so $(K,+,\cdot,\tau')$ is also a KT-near-field, i.e. (L,K) is a KT-near-field extension.

Vice versa if (L,K) is a KT-near-field extension and $\overline{L} := L \cup \{\infty\}$, $\overline{K} := K \cup \{\infty\}$ then the permutations

$$a^+: \overline{L} \to \overline{L};\ x \to a+x \text{ for } x \in L;\ \infty \to \infty \text{ for all } a \in L,$$

$$a^\cdot: \overline{L} \to \overline{L};\ x \to a\cdot x \text{ for } x \in L;\ \infty \to \infty \text{ for all } a \in L^*$$

$$\sigma: \overline{L} \to \overline{L};\ x \to \tau(x) \text{ for } x \in L^*;\ 0 \to \infty,\ \infty \to 0$$

generate a group Γ which acts sharply 3-transitive on the set $\overline{L}$, and if we put $\mathfrak{K} := \Gamma(\overline{K})$ then $(\overline{L},\mathfrak{K})$ is a Möbius structure of type $S(3,|K|+1)$ and Γ is a subgroup of the automorphism group of $(\overline{L},\mathfrak{K})$. Furthermore the derivation in each point is an affine space.

Remarks. 1. Each finite Desarguesian affine space $(P,\mathfrak{L})$ can be embedded in a Möbius structure $(\overline{P},\mathfrak{K})$ such that $\overline{P} = P \dot\cup \{\infty\}$ and $\overline{\mathfrak{L}} := \{L \dot\cup \{\infty\} \mid L \in \mathfrak{L}\} \subset \mathfrak{K}$, i.e. $(P,\mathfrak{L})$ is the derivation of $(\overline{P},\mathfrak{K})$ in the point ∞, and such that $(\overline{P},\mathfrak{K})$ admits an automorphism group Γ which acts sharply 3-transitive on the point set $\overline{P}$: Since $(P,\mathfrak{L})$ is desarguesian there is a finite field K such that $(P,\mathfrak{L})$ has the representation $P = K^n$ and $\mathfrak{L} = \{\mathfrak{a}+K\mathfrak{b} \mid \mathfrak{a},\mathfrak{b} \in K^n,\ \mathfrak{b} \neq (0,\dots,0)\}$. Furthermore we know that there is a finite field extension L of K with $[L:K] = n$, i.e. K^n can be identified with L. Since $(L,+,\cdot)$ is a commutative field, the map $\tau: L^* \to L^*;\ x \to x^{-1}$ satisfies the functional equation $(*)$, and so $(L,+,\cdot,\tau)$ is a KT-near-field containing the KT-near-field $(K,+,\cdot,\tau')$ with $\tau' := \tau|K^*$.

2. W. BENZ ([23], §7) characterized Möbius planes which admit a sharply 3-transitive automorphism group by quadratic KT-near-field extensions. These Möbius planes are Miquelian (cf. [4] p. 206, 229 Satz 3.2) if and only if the quadratic KT-field extension (L,K) consists of two commutative fields.

12. Laguerre Structures and Laguerre Codes

Let $(P,\mathfrak{G},\mathfrak{K})$ be a 1-chain structure, let again $k := |\mathfrak{G}|$ and $q := |E|$ for $E \in \mathfrak{G}$. Then by **§4**, $\mathfrak{K} \subset K^k$ can be considered as a q-nary code of length k where K is an alphabet with $|K| = q$. If there is a number $t \in \mathbb{N}$ such that for any t-set $\{x_1,x_2,\dots,x_t\} \in \binom{P}{t}$ with $x_i \frown x_j$ for $i,j \in \{1,2,\dots,t\}$ with $i \neq j$ there is exactly one chain $C \in \mathfrak{K}$ with $\{x_1,x_2,\dots,x_t\} \subset C$, then W. HEISE [1] calls such a structure an **optimal (k,t)-geometry of order q**. From this definition it follows $|\mathfrak{K}| = q^t$. HEISE showed that these geometries correspond with the optimal (k,t)-codes $\mathfrak{K} \subset K^k$. For $t=2$ these codes are the (MDS)-codes which we studied in the sections 8 and 10.

Here we will consider the case $t=3$, i.e. $(P,\mathfrak{G},\mathfrak{K})$ is a Laguerre structure and so $|\mathfrak{K}| = q^3$. The corresponding code we call a **Laguerre-code.** To avoid trivial cases we assume $k > 3$. Then the derivation $(P,\mathfrak{G},\mathfrak{K})^{(p)}$ in a point $p \in P$ is a 1-structure with the parameters $k-1$ and q, and a 1-chain structure $(P,\mathfrak{G},\mathfrak{K})$ is a Laguerre structure if and only if the derivation in each point $a \in P$ is a 1-structure. By **(3.5)** $k-1 \le q+1$ hence $k \le q+2$.

There are Laguerre structures with $k = q+2$. For the smallest possible values $q = 2$, $k = 4$ there is exactly one such structure: Let $P = \mathbb{Z}_2^3$ be the set of points of the 3-dimensional affine space $A(\mathbb{Z}_2^3,\mathbb{Z}_2)$ over the prime field $\mathbb{Z}_2$ of characteristic 2, let $\mathfrak{G}$ be a pencil of parallel lines of $A(\mathbb{Z}_2^3,\mathbb{Z}_2)$ and let $\mathfrak{K}$ be the set of all planes which are not parallel to one line of $\mathfrak{G}$. Then $(P,\mathfrak{G},\mathfrak{K})$ is the Laguerre structure and we have $v := |P| = |\mathfrak{K}| =: b = 8$. This follows from the fact that each plane of $\mathfrak{K}$ intersects each line of $\mathfrak{G}$ in exactly one point and that any three points a,b,c with $[a] \neq [b] \neq [c] \neq [a]$ are not collinear and so determine exactly one plane of $\mathfrak{K}$.

This construction can be carried out in each 3-dimensional affine space $A(K^3,K)$ over a finite field of characteristic 2 in the following way:
Since K is finite there is an $\alpha \in K$ such that the polynomial $x + \alpha x + 1$ is irreducible. Then the equation $x^2 + \alpha xy + y^2 = 1$ defines an ellipse O in the plane $z=0$ where the point $(0,0,0)$ is the knot of O, i.e. each line of the plane $z=0$ through $(0,0,0)$ meets O in exactly one point (this is the case since $\mathrm{char}(K) = 2$). Therefore $|O| = |K|+1$.

Let Z be the z-axis $(x = y = 0)$, let $\mathfrak{G} := \{Z\} \cup \{(p \parallel Z) \mid p \in O\}$ be the set of all lines parallel to Z which meet O or (0,0,0), and let $P := \cup \mathfrak{G}$ be the set of all points which are lying on one of the lines of $\mathfrak{G}$. Finally let $\mathfrak{E}'$ be the set of all planes which intersect the line Z in exactly one point and let $\mathfrak{K} := \{E \cap P \mid E \in \mathfrak{E}'\}$. Then $(P, \mathfrak{G}, \mathfrak{K})$ is a Laguerre structure with the parameters $q := |K|$, $k = q+2 = |O|+1$, $v := |P| = q \cdot k = q(q+2)$, $b := |\mathfrak{K}| = q^3$.

(12.1) Let $(P, \mathfrak{G}, \mathfrak{K})$ be a finite Laguerre structure with the parameters $k = q+2$. Then q and so k are even.

Proof. Let $p \in P$; then the derivation $(P', \mathfrak{G}', \mathfrak{K}') := (P, \mathfrak{G}, \mathfrak{K})^P$ in the point p has the parameters $q' = q$ and $k' = q'+1 = q+1$ and $(P', \mathfrak{G}' \cup \mathfrak{K}')$ is by §8 (cf. section: (MDS)codes and Parallelisms) a punctured projective plane with $\mathfrak{L}_0 = \mathfrak{K}'$ and $\mathfrak{L}_1 = \mathfrak{G}'$. Therefore any two distinct chains $A, B \in \mathfrak{K}'$ intersect in exactly one point. But this implies for our Laguerre structure $(P, \mathfrak{G}, \mathfrak{K})$ that the intersection $X \cap Y$ of two distinct chains $X, Y \in \mathfrak{K}$ is either empty or consists of exactly two points. Now let $C \in \mathfrak{K}$, $a, b \in P \setminus C$ with $[a] \neq [b]$ and $a' := [a] \cap C$, $b' := [b] \cap C$. Then to each $x \in C \setminus \{a', b'\}$ there is exactly one chain $X \in \mathfrak{K}$ with $a, b, x \in X$ and hence $|C \cap X| = 2$. This showes that $q = k-2 = |C \setminus \{a', b'\}|$ is even.

A Laguerre structure $(P, \mathfrak{G}, \mathfrak{K})$ is a finite Laguerre plane of order q if $k = q+1$. The touching-axiom **T** (cf. §3) which we have to claim in the infinite case, can be proved in the finite case: Let $A \in \mathfrak{K}$, $a \in A$ and $b \in P \setminus (A \cup [a])$. Then $a \frown b$. Since t =3 and $|K| = q$, $\mathfrak{K}(a,b) := \{X \in \mathfrak{K} \mid a, b \in X\}$ consists of q chains and to each point $x \in A \setminus (\{a\} \cup [b])$ there is exactly one chain $X \in \mathfrak{K}(a,b)$ with $x \in X$. Hence $q = |\mathfrak{K}(a,b)| \geq |A \setminus (\{a\} \cup [b])| = |A \setminus (\{a\} \cup (A \cap [b]))| = k-2 = q-1$, and so there is exactly one chain $B \in \mathfrak{K}(a,b)$ with $B \cap A = \{a\}$. This shows us that every Laguerre code $\mathfrak{K} \subset K^k$ determines a Laguerre plane if $k = |K|+1 = q+1$.

To every finite field K there corresponds a finite Miquelian Laguerre plane $(P, \mathfrak{G}, \mathfrak{K})$ (cf. [4] p. 131, Satz 3.4 and p. 93 pp) which can be constructed as above by forming a cylinder over an ellipse in the 3-dimensional affine space over K. (This **cylinder model** was first introduced 1911 by W. BLASCHKE in the classical Euclidean space (cf. [24])).

There is also an algebraic representation of the Laguerre codes corresponding to Miquelian Laguerre planes which we obtain from the algebraic presentation of these planes by the dual numbers of E. STUDY (cf. [3] p. 107):

(**12.2**) Let K be a finite field, $\overline{K} := K \cup \{\infty\}$ and for $a,b,c \in K$ let

$$C_{a,b,c}: \begin{cases} \overline{K} \to K \\ \infty \to a \\ x \to -ax^2 + (a+c-b)x + b \text{ for } x \in K \end{cases}$$

Then $\mathfrak{C} := \{C_{a,b,c} \mid a,b,c \in K\}$ is a q-nary Laguerre code with $q := |K|$ of length $k := |\overline{K}| = |K| + 1$. Furthermore $\mathfrak{C}$ is linear, and if $K = \{0,1,t_3,\dots,t_q\}$ then $C_{011} = (0,1,1,\dots,1)$, $C_{001} = (0,0,1,t_3,t_4,\dots,t_q)$ and $C_{-101} = \left(-1,0,1,t_3^2, t_4^2, \dots, t_q^2\right)$ form a basis of $\mathfrak{K}$.

In the case of char(K) = 2 we obtain by forming the cylinder over an ellipse and its knot the following algebraic representation:

(**12.3**) Let K be an finite field with char(K) = 2, let $\alpha \in K$ such that the polynomial $t^2 + \alpha t + 1$ is irreducible, let $q := |K|$ and let $\overline{\overline{K}} := K \mathbin{\dot\cup} \{\infty_1\} \mathbin{\dot\cup} \{\infty_2\}$.
For $a,b,c \in K$ let

$$C_{a,b,c}: \overline{\overline{K}} \to K;\ \infty_1 \to a;\ \infty_2 \to \frac{b+a}{\alpha} + c;$$

$$x \to \frac{(\alpha(a+c) + (a+b))x^2 + \alpha(b+c) + (a+b)}{\alpha(1+\alpha x + x^2)} + \frac{b+a+\alpha c}{\alpha} \text{ for } x \in K.$$

Then $\mathfrak{C} := \{C_{a,b,c} \mid a,b,c \in K\}$ is a q-nary Laguerre code of length $k = |\overline{\overline{K}}| = q + 2$.

13. 2-Structures, Hyperbola Structures, Minkowski Planes and Corresponding Codes

Here we consider 2-chain structures $(P, \mathfrak{G}_1, \mathfrak{G}_2, \mathfrak{K})$. Then for all $A \in \mathfrak{G}_1 \cup \mathfrak{G}_2$ and $X \in \mathfrak{K}$, $|A \cap X| = 1$. This implies (cf. (3.2)) that there is a cardinal number k with $|X| = k$ for all $X \in \mathfrak{G}_1 \cup \mathfrak{G}_2 \cup \mathfrak{K}$, $|\mathfrak{G}_1| = |\mathfrak{G}_2| = k$ and $v := |P| = k^2$. Furthermore we can formulate the two **rectangle axioms** (cf. [5], [6] p. 64, 65):

R_S For $A,B,C \in \mathfrak{K}$ the set $\{[[a]_1 \cap B]_2 \cap [[a]_2 \cap C]_1 \mid a \in A\}$ is a chain of $\mathfrak{K}$.

$R_{E\,i}$ For $E \in \mathfrak{K}$ and any $B \in \mathfrak{K}$ the set $\{[[[a]_1 \cap B]_2 \cap E]_1 \cap [a]_2 \mid a \in E\}$ is a chain of $\mathfrak{K}$.

and the **symmetry axiom** (cf.[4] p. 302, [25] p. 84, [6] p. 75):

S For each $A \in \mathfrak{K}$ let $\hat{A}: P \to P;\ x \to [[x]_1 \cap A]_2 \cap [[x]_2 \cap A]_1$; then $\hat{A}(X) \in \mathfrak{K}$ for all $X \in \mathfrak{K}$.

If $(P, \mathfrak{G}_1, \mathfrak{G}_2, \mathfrak{K})$ is even a 2-structure and $A, B \in \mathfrak{G}_1$ with $A \neq B$ and $a \in A$, then to each $b \in B \setminus [a]_2$ there is exactly one $X \in \mathfrak{K}$ with $a, b \in X$. Therefore $|\mathfrak{K}| = k(k-1)$. In the case that $(P, \mathfrak{G}_1, \mathfrak{G}_2, \mathfrak{K})$ is a hyperbola structure, the derivation in a point $p \in P$ is a 2-structure with the parameter $k' = k-1$ (cf. §3). This gives us here the result $|\mathfrak{K}| = k(k-1)(k-2)$.

By (3.3) to every 2-chain structure there corresponds the 1-chain structure $(P, \mathfrak{G}_1, \mathfrak{C} := \mathfrak{G}_2 \cup \mathfrak{K})$. Therefore we can associate to any finite 2-chain structure the following codes:

(13.1) Let $(P, \mathfrak{G}_1, \mathfrak{G}_2, \mathfrak{K})$ be a finite 2-chain structure with the parameter k, let $E \in \mathfrak{K}$, $K \in \mathfrak{G}_1$ and $K_2 \in \mathfrak{G}_2$ be distinct and let $\iota: K \to K_2;\ x \to [[x]_2 \cap E]_1 \cap K_2$. Each $G \in \mathfrak{G}_2$ we identify with the constant map $G: K_2 \to K;\ x \to G \cap K$, each $C \in \mathfrak{K}$ with the map $C: K_2 \to K;\ x \to [[x]_1 \cap C]_2 \cap K$ and we set $\tilde{C} := C \circ \iota$. Then:

a) $\mathfrak{K}$ and $\mathfrak{C} := \mathfrak{G}_2 \cup \mathfrak{K}$ are k-nary codes of length k where $\mathfrak{K}$ consists of bijections and $\tilde{\mathfrak{K}} := \{\tilde{C} \mid C \in \mathfrak{K}\}$ of permutations. Therefore we will identify the set $\mathbb{Z}_k = K_2$ with the alphabet K, and so $\mathfrak{K}$ and $\tilde{\mathfrak{K}}$.

b) $\tilde{\mathfrak{K}}$ is a semigroup if and only if the rectangle axiom $\mathbf{R}_S$ is valid, and $\tilde{\mathfrak{K}}$ is a group if and only if $\mathbf{R}_S$ and $\mathbf{R}_{Ei}$ are valid (cf. [5], [6]).

c) The following statements are equivalent:

α) $(P, \mathfrak{G}_1, \mathfrak{G}_2, \mathfrak{K})$ is a 2-structure.

β) $\mathfrak{C}$ is an (MDS)-code.

γ) $(K, \tilde{\mathfrak{K}})$ is a **sharply 2-transitive permutation set,** i.e. $\forall (a,b), (c,d) \in K^2$ with $a \neq b$, $c \neq d$, $\exists_1\ \tilde{C} \in \tilde{\mathfrak{K}}$ with $\tilde{C}(a) = c \wedge \tilde{C}(b) = d$.

d) $(P, \mathfrak{G}_1, \mathfrak{G}_2, \mathfrak{K})$ is a 2-structure with rectangle axiom $\mathbf{R}_S$ if and only if there is a finite (right) nearfield $(K, +, \cdot)$ such that $\tilde{\mathfrak{K}} = \{a^+ b^\cdot \mid a, b \in K,\ b \neq 0\}$ where $a^+: K \to K;\ x \to a + x$ and $b^\cdot: K \to K;\ x \to x \cdot b$ (i.e. $\tilde{\mathfrak{K}}$ is the affine group of the nearfield $(K, +, \cdot)$) (cf. [5]).

e) The following statements are equivalent.

α) $(P, \mathfrak{G}_1, \mathfrak{G}_2, \mathfrak{K})$ is a Minkowski plane.

β) $(K, \tilde{\mathfrak{K}})$ is a sharply 3-transitive permutation set.

γ) $\mathfrak{K}$ is a **Minkowski code,** i.e. $\forall h,i,j \in K_2$ $(= \mathbf{Z}_k)$ with $\{h,i,j\} \in \binom{K_2}{3}$ and $\forall \alpha,\beta,\gamma \in K$ with $\{\alpha,\beta,\gamma\} \in \binom{K}{3}$ $\exists_1$ $C \in \mathfrak{K}$ such that $C(h) = \alpha$, $C(i) = \beta$, and $C(j) = \gamma$.

f) For a Minkowski plane the following statements **α), β), γ)** (cf. [25] §5), resp. **α'), β')** (cf.[26], [27]) are equivalent.

α) The rectangle axiom $\mathbf{R}_S$

β) $\tilde{\mathfrak{K}}$ is a group.

γ) There is a finite **K.T. nearfield** $(N, +, \cdot, \sigma)$, (i.e. $(N, +, \cdot)$ is a nearfield and σ is an involutory automorphism of the multiplication group $(N^*, \cdot)$ such that: $\forall x \in N\setminus\{0,-1\}$: $\sigma(1+\sigma(x)) = 1-\sigma(1+x)$.) with $|N| = k-1$ such that the Minkowski code $\mathfrak{K}$ is the permutation group of the set

$K = N \dot{\cup} \{\infty\}$ generated by the permutations:

$a^+(x) = a+x$ for $x \in N$ and $a^+(\infty) = \infty$; $a \in N$

$b^{\cdot}(x) = x \cdot b$ for $x \in N$ and $b^{\cdot}(\infty) = \infty$; $b \in N^* := N\setminus\{0\}$

$\bar{\sigma}(x) = \sigma(x)$ for $x \in N^*$, $\bar{\sigma}(0) = \infty$ and $\bar{\sigma}(\infty) = 0$.

α') The symmetry axiom **S**.

β') There is a finite field F of order $k-1$ such that the Minkowski code $\mathfrak{K}$ consists of the permutations $C_{a,b,c,d}$ with $a,b,c\,d \in K$ and $ad \neq bc$ of the set $K = F \dot{\cup} \{\infty\}$ defined by $x \to \frac{ax+b}{cx+d}$ for $x \in F\setminus\{-\frac{d}{c}\}$, $-\frac{d}{c} \to \infty$, $\infty \to \frac{a}{c}$ if $c \neq 0$, and $x \to \frac{ax+b}{d}$, $\infty \to \infty$ if $c = 0$.

Remarks. 1. For a hyperbola structure the symmetry axiom **S** implies the rectangular axioms $\mathbf{R}_S$ and $\mathbf{R}_{Ei}$ (cf. [26], [27]) and the touching axiom **T** (cf. [25]). Therefore the axiom **S** characterizes the Miquelian Minkowski planes (cf. [4]). But there are even finite Minkowski planes with $\mathbf{R}_S$ and $\mathbf{R}_{EI}$ where **S** is not valid. This follows from the existence of finite K.T.-nearfields which are not fields (cf. e.g. [28]).

2. While for a finite 2-structure $(P, \mathfrak{G}_1, \mathfrak{G}_2, \mathfrak{K})$, $\mathfrak{C} := \mathfrak{G}_2 \cup \mathfrak{K}$ is an (MDS)-code we do not obtain in this sense a Laguerre code if $(P, \mathfrak{G}_1, \mathfrak{G}_2, \mathfrak{K})$ is a finite Minkowski plane: Let $\alpha,\beta \in K$ be distinct and let $h,i,j \in K$ with $\{h,i,j\} \in \binom{K}{3}$; then there is no function $C \in \mathfrak{K}$ with $C(h) = C(i) = \alpha$ and exactly one function

$G \in \mathfrak{G}_2$ with $G(h) = G(i) = \alpha$, but then $G(j) = \alpha \neq \beta$.

Acknowledgements. I have to thank firstly Professor Mario Marchi from the University of Udine, Italy (who made me work and learn in order to prepare these lectures) and the CISM for the great honour to invite me to be one of the lecturers at the Advanced School on "Geometries, Codes and Cryptography" in Udine; secondly Dr. Alan Oswald from the Teesside Polytechnic, England (who offered me the position of a visiting senior lecturer for the months March and April 1989 with the only obligation, to do research) for his cooperation in preparing these lecturers. During my stay in Middlesbrough we had every week several discussions and many results of this paper were discovered together with Dr. Oswald. Thirdly I have to thank Mrs. Jutta Hilspach for typing, and Dr. Alexander Kreuzer and Hubert Kiechle for making up into pages the typescript on the word processing system.

REFERENCES

[1] HALDER, H.-R., HEISE, W.: Einführung in die Kombinatorik. Carl Hanser Verlag, München Wien 1976

[2] HEISE, W., QUATTROCCHI, P.: Informations- und Codierungstheorie. Springer Verlag, Berlin-Heidelberg-New York 1989

[3] KARZEL, H., KROLL, H.-J.: Geschichte der Geometrie seit Hilbert. Wiss. Buchgesellschaft Darmstadt 1988

[4] BENZ, W.: Vorlesung über Geometrie der Algebren. Berlin-Heidelberg- New York 1973

[5] KARZEL, H.: Zusammenhänge zwischen Fastbereichen, scharf zweifach transitiven Permutationsgruppen und 2-Strukturern mit Rechtecksaxiom. Abh. Math. Sem. Univ. Hamburg **32** (1968), 191-206

[6] KARZEL, H., KROLL, H.-J.: Perspectivities in circle geometries. In: Geometry - von STAUDT'S point of View. Nato Advanced Study Institutes Series C: Math. and Phys. Sci. Dordrecht, Holland (1981), 51-99

[7] KARZEL, H., SÖRENSEN, K., and WINDELBERG, D.: Einführung in die Geometrie. Göttingen 1973

[8] KARZEL, H.: Kinematic Spaces. Symp. Matematica, Ist. Naz. di Alta Matematica **11** (1973), 413-439

[9] KIST, G.: Projektiver Abschluß 2-gelochter Räume. Res. Math. **3** (1980), 192-211

[10] SÖRENSEN, K.: Ein Beweis von H. KARZEL und I. PIEPER. J. of Geometry **32** (1988), 131-132

[11] KARZEL, H.: Porous Double Spaces. J. of Geometry **34** (1989), 80-104

[12] BUEKENHOUT, F.: Une caracterisation des espaces affins basee sur la notion de droite. Math. Z. **111** (1969), 367-371

[13] KARZEL, H., PIEPER, I.: Bericht über geschlitzte Inzidenzgruppen. Jber. DMV **72** (1970), 70-114

[14] KLOSSEK, S.: Kommutative Spiegelungsräume. Mitt. Math. Sem. Gießen **117** (1975)

[15] SOUBLIN, J.P.: Etude algebrique de la notion de moyenne. Journal de Mathematiques pures et appliques **50** (1971)

[16] KARZEL, H., OSWALD, A.: Near-rings, (MDS)- and Laguerre Codes. To appar

[17] PILZ, G.: Near-rings. Amsterdam-NewYork-Oxford 1977

[18] MELDRUM, J.: Near-rings and their links with groups. Pitman Publ. Co. (Research Note Series No. **134**), 1985

[19] WÄHLING, H.: Theorie der Fastkörper. Thales Verlag Essen 1987

[20] WITT, E.: Die 5-fach transitiven Gruppen von Mathieu. Abh. Math. Sem. Univ. Hamburg **12** (1938), 256-264

[21] HEISE, W.: Eine Definition des Möbiusraumes. Manuscripta math. **2** (1970), 39-47

[22] KERBY, W.: On infinite sharply multiply transitive groups. Hamburger Math. Einzelschr., Neue Folge **6**, Göttingen 1974

[23] BENZ, W.: Über Möbiusebenen. Ein Bericht. Jber. DMV **63** (1960), 1-27

[24] BLASCHKE, W.: Über die Laguerre'sche Geometrie orientierter Geraden in der Ebene I. In: BLASCHKE, W. Gesammelrte Werke, Band 1. Thales Verlag Essen (1982), 253-261

[25] HEISE, W., KARZEL, H.: Symmetrische Minkowski-Ebenen. J. of Geometry **3** (1973), 5-20

[26] ARTZY, R.: A Pascal Theorem Applied to Minkowski Geometry. J. of Geometry **3** (1973), 93-105

[27] KARZEL, H.: Symmetrische Permutationsmengen. Äquationes Mathematicae **17** (1978), 83-90

[28] WEFELSCHEID, H.: Verallgemeinerte Minkowski-Geometrie. In: Contributions to Geometry. Proceedings of the Geometry Symposium in Siegen 1978. Edited by J. Tölke and J. Wills. Birkhäuser Verlag Basel 1979

TOPICS IN ALGEBRAIC CODING THEORY

W. Heise
Technical University, München, F.R.G.

Abstract

This article surveys some selected topics in algebraic coding theory and their links to geometry and cryptography.

In the first chapter the model of a discrete communication system with a noisy memoryless and stationary symmetric channel is described. It is shown, how memorylessness can approximately be achieved by the method of interleaving. To guarantee the security of data transmission through this noisy channel one uses an encoder, which concatenates every word $\mathbf{m}$ consisting of k information symbols with a word $\mathbf{r}$ consisting of r redundant symbols. Passing through the channel some of the $n := k+r$ symbols of the codeword $\mathbf{c} := (\mathbf{m},\mathbf{r})$ will eventually be changed into other symbols. The maximum-likelihood-decoder receives a word $\mathbf{y}$ and tries to recover $\mathbf{c}$ by searching an admissible codeword, which differs from $\mathbf{y}$ in a minimum number of symbols. A good coding system uses a code with a high information rate k/n but which nevertheless is capable to correct the expected number of symbol errors. Some coding bounds show, which compromises one has to accept, when designing an optimal coding system. The implementation of the encoder and decoder is facilitated, if the used code bears a mathematical structure, i. e. if it is linear or a forteriori cyclic.

The second chapter deals with some important classes of codes like Goppa-codes, Hamming- and Simplex-codes, MDS-codes and Reed-Muller-codes. In one section we show the connection between MDS-codes and Laguerre-geometry, another section describes the automorphism group of general linear codes.

The third chapter studies more en detail the quadratic residue codes.

1. Basic Concepts

1.1. The q-nary Symmetric Channel

Idealizing we assume that our noisy discrete communication system is equipped with a q-**nary symmetric channel** "qSC" of *error probability*

$$p < (q-1)/q :$$

The qSC accepts at its input at every tick of a clock, say every "second", one symbol α from a discrete q-element *alphabet* F. Mostly we assume q to be a prime power and impose on F the structure of the finite Galois field GF(q). After having passed the channel the input symbol α will be emitted without damage with the probability $1-p$. But with the probability p an output symbol $\beta \in F$ different from α will be emitted. The a priori probability that after the input of a symbol α a prescribed symbol $\beta \in F\setminus\{\alpha\}$ will be emitted does not depend on β, i. e.

$$p(\beta|\alpha) = p/(q-1) \quad \text{for} \quad \beta \neq \alpha .$$

The qSC is *memoryless* and *stationary*, i. e. the errors occur independently from each other and do not depend on the concrete time instant.

1.2. Interleaving

In reality errors tend to occur in bursts; but with a simple trick we can achieve memorylessness approximately : Before entering the channel the incoming symbols are linewise written into a matrix shaped memory, the so-called **interleaver**. When the interleaver is full the symbols will columnwise be read by the channel entrance. After the channel output the symbols will columnwise be read into a **deinterleaver** and linewise be read by the following device. With this procedure burst errors will be split into independent errors. The size of the memory matrices depends on the length of the expected bursts. This interleaving causes only a constant delay in the data transmission of about as many seconds as there are memory cells in the interleaving devices.

1. 3. Code Implementation

To combat errors we encode the data. The sender emits every n seconds one message chosen from a source S consisting of a finite number of messages. We often assume that the admissible messages are all the q^k different words of length k with components in F, i. e. $S := F^k$. The theory of data compression suggests to implement devices into the sender, which transform the original message source into this form. But even if the source does not have this form, the *entropy* of the source S, i. e. its mean information content, measured in q-nary information units, has the value

$$k := \log_q |S|.$$

The **encoder** translates every message using an injective coding map $C : S \to F^n$ into a **codeword** $\mathbf{c} = c_0 c_1 \ldots c_{n-1}$ of the **block code** $\mathbf{C} := C(S) \subseteq F^n$. The number n is called the **block length** of $\mathbf{C}$. Passing the qSC some components of $\mathbf{c}$ may be changed into other symbols and a word $\mathbf{y} = y_0 y_1 \ldots y_{n-1} \in F^n$ will be emitted. The **maximum likelihood decoder** MLD looks in a dictionary of all codewords for a codeword in $\mathbf{C}$ nearest to $\mathbf{y}$, i. e. it picks up a codeword $\mathbf{d} \in \mathbf{C}$ with minimum **Hamming distance**

$$\varrho := \varrho(\mathbf{d},\mathbf{y})$$

from $\mathbf{y}$, i. e. it picks up a codeword $\mathbf{d} \in \mathbf{C}$, which differs from $\mathbf{y}$ in a minimum number ϱ of components. Then it sends the message $C^{-1}(\mathbf{d})$ to the receiver.

1. 4. Correction and Information Rate

If there happen t errors during the transmission of the codeword $\mathbf{c} \in \mathbf{C}$, i. e . if the Hamming distance between the channel input word $\mathbf{c} \in \mathbf{C}$ and the channel output word $\mathbf{y} = \in F^n$ has the value $\varrho(\mathbf{c},\mathbf{y}) = t$, and if there is no codeword $\mathbf{x} \in \mathbf{C}$ with $\varrho(\mathbf{c},\mathbf{x}) \le 2t$, then the MLD will correct the t errors, the codewords $\mathbf{c}$ and $\mathbf{d}$ coincide.

In general : If we use a code $\mathbf{C} \subseteq F^n$ with **minimum distance**

$$d := d(\mathbf{C}) := \min\{\varrho(\mathbf{c},\mathbf{d});\ \mathbf{c},\mathbf{d} \in \mathbf{C}, \mathbf{c} \neq \mathbf{d}\},$$

then the MLD is able to correct any error pattern of up to $(d-1)/2$ errors per codeword, no matter which codeword was transmitted. The ratio $\lambda := d/n$ is called the **correction rate** of $\mathbf{C}$. Since we expect on the average $p \cdot n$ errors per codeword transmitted over the qSC, the correction rate λ of $\mathbf{C}$ should exceed the double value of the error probability p of the channel:

$$2p < \lambda = d/n.$$

The entropy of the source S, i. e. the parameter

$$k := \log_q|S| = \log_q|\mathbf{C}|$$

is also called the **dimension** of the code $\mathbf{C}$. Measured in q-nary information units, the **information rate** $R := k/n$ of $\mathbf{C}$ indicates the mean content of information transported through the channel by symbol.

If we have the possibility to choose between two codes of different lengths but of the same correction and information rate, then as a consequence of Čebyšev's weak law of the large numbers we should prefer the longer code: The chance of occurrence of considerably more than $\lambda \cdot n/2$ errors for one transmitted codeword tends to zero with increasing length n, and so does the probability of a decoding failure. Unfortunately until yet nobody found a series of practicable codes of increasing block lengths with reasonable constant correction and information rate. This evil fact is known as the *Empiric Law on the Coding Theorists' Mental Limitations.*

1.5. Coding Bounds

A good error correcting code has a high length n, a high minimum distance d and a high information rate R. Of course, we cannot get altogether: Observing that the closed balls of radius $(d-1)/2$ with respect to the Hamming metric centered around the q^k codewords must be disjoint, one gets with the **Hamming bound**

$$\sum_{i=0}^{(d-1)/2} \binom{n}{i}(q-1)^i \le q^{n-k}$$

an upper bound of the information rate in dependence of the minimum distance. Codes, for which the Hamming bound holds with equality are called **perfect**.

The minimum distance cannot exceed the mean distance between two codewords. This fact implies the **Plotkin bound**

$$\lambda \le 1 - \frac{1}{q}\left(\frac{1}{1-q^{-k}}\right),$$

which holds with equality if and only if the code is *equidistant* and *equidistributed*, i. e. if any two distinct codewords have the same Hamming distance d and if every symbol of F occurs in every position of exactly q^{k-1} codewords.

Observing that the q^k prefixes of length $n-d+1$ of the codewords must be pairwise different, we obtain the **Singleton bound**

$$d \leq n-k+1,$$

which does not depend on the alphabet size q. The Singleton bound holds with equality, if and only if the code $\mathbf{C}$ is **maximum distance separable** (MDS), i. e. if for any k indices $i_1 < i_2 < \ldots < i_k$ and any k symbols $\alpha_1, \alpha_2, \ldots, \alpha_k \in F$ there is exactly one codeword $\mathbf{c} = c_0 c_1 \ldots c_{n-1} \in \mathbf{C}$ with $c_{i_j} = \alpha_j$ for $j = 1, 2, \ldots, k$.

1. 6. Linear Codes

The perfect codes, the equidistant and equidistributed codes and the MDS codes are **optimal** codes, i. e. codes for which there do not exist other codes over the same alphabet of the same length n and the same minimum distance d but with a higher information rate. In case that q is a prime power, code designers because of their easier implementation prefer **linear** (n,k)**-codes**, that are k-dimensional linear subspaces of the vector space F^n of all n-tuples with components in $F = GF(q)$. It turns out, that for most optimal codes over alphabets of prime power size there exist linear codes with the same parameters. Thus from an engineering point of view the restriction to linear codes is not a very serious handicap. The advantage of linear codes lies in the fact that many code manipulations can be performed by calculations instead of time wasting table lookups.

We can represent any codeword of a linear (n,k)-code $\mathbf{C} \subseteq F^n$ as a linear combination of the rows of a **generator matrix** G of $\mathbf{C}$, that is a $k \times n$-matrix, whose rows form a basis of the linear space $\mathbf{C}$. We put $r := n-k$. A generator matrix H of the **dual code** $\mathbf{C}^\perp$ of $\mathbf{C}$, i. e. the r-dimensional subspace orthogonal to $\mathbf{C}$ with respect to the ordinary scalar product in F^n, is called a **check matrix** of $\mathbf{C}$. Defining by $H \cdot \mathbf{y} \in F^r$ the **syndrome** of a word $\mathbf{y} \in F^n$, we can use the check matrix for accelerating the maximum likelihood decoder : The Hamming metric on F^n is invariant against translations, and thus the distances between any two vectors in F^n are all known, if we only know the **Hamming weight** $\gamma(\mathbf{x}) := \varrho(\mathbf{0}, \mathbf{x})$ for all vectors $\mathbf{x} \in F^n$. Now, if in the qSC an error vector $\mathbf{e} \in F^n$ is added to a codeword $\mathbf{c} \in \mathbf{C}$, then the MLD calculates the syndrome $\mathbf{s} := H\cdot\mathbf{y}$ of the received word $\mathbf{y} := \mathbf{c}+\mathbf{e}$. Knowing that $\mathbf{e}$ belongs to the coset $\mathbf{y}+\mathbf{C}$ and assuming that the unknown error vector $\mathbf{e}$ has minimal weight in $\mathbf{y}+\mathbf{C}$, the MLD picks up from a list the **coset leader** $\mathbf{f} \in \mathbf{y}+\mathbf{C}$, i. e. a predetermined vector of minimal Hamming weight in $\mathbf{y}+\mathbf{C}$. In this list the q^r cosets are identified by their syndromes.

1. 7. Cyclic Codes

Although much faster than a pure table pickup algorithm the syndrome decoding procedure is too complex for practical purposes. There exist more powerful coding and decoding techniques for codes with a richer mathematical structure. A linear (n,k)-code $\mathbf{C} \subseteq F^n$ is said to be **cyclic**, if for every codeword $\mathbf{c} = c_0c_1 \ldots c_{n-2}c_{n-1} \in \mathbf{C}$ the cyclic shifted word $c_{n-1}c_0c_1 \ldots c_{n-2}$ is again a codeword. We impose now on F^n the structure of the residue class algebra

$$R_n(q) := F[z]/(z^n-1),$$

in which we always calculate with representatives of minimal degree. Identifying any vector $\mathbf{x} = x_0x_1 \ldots x_{n-2}x_{n-1} \in F^n$ with its generating function

$$x(z) := \sum_{i=0}^{n-1} x_i z^i \in R_n(q)$$

we see, that the cyclic (n,k)-codes over F coincide with the k-dimensional ideals in $R_n(q)$.

The **generator polynomial** $g(z)$ of a cyclic (n,k)-code $\mathbf{C} \subseteq R_n(q)$ is defined as the monic polynomial of minimal degree in $\mathbf{C}$. It is a divisor of the polynomial z^n-1 and has degree $\deg g(z) = r := n-k$. The reversion of the **check polynomial** $h(z) := (z^n-1)/g(z)$ of $\mathbf{C}$ generates in $R_n(q)$ as an ideal the dual code $\mathbf{C}^\perp$ of $\mathbf{C}$.

The algebra $R_n(q)$ is semisimple if and only if q and n are relatively prime. Mathematical difficulties in dealing with non semisimple algebras led coding theorists to stress on cyclic codes for which the characteristic p of F does not divide the length n, although there exist interesting cyclic codes in the other case : Since $z^p-1 = (z-1)^p$ over $GF(p)$ all the $p+1$ ideals in $R_p(p)$ are MDS codes. In case that $R_n(q)$ is semisimple any cyclic code $\mathbf{C}$ contains a unit element $i(z)$ which is called somewhat misleadingly *the* **idempotent** of $\mathbf{C}$.

The calculation of the coefficients of the generator polynomial of a cyclic code $\mathbf{C} \subseteq R_n(q)$ over a field $F := GF(q)$ of characteristic p may be a tedious task. Factorizing z^n-1 in cyclotomic polynomials can be of great help, but the cyclotomic polynomials over F are in general not irreducible. Let v be the largest power of p dividing n and set $t := n/v$. Then $z^n-1 = (z^t-1)^v$. Denote by ζ a primitive t^{th} root of unity over F. Defining for $j = 0,\ldots,t-1$ the **cyclotomic coset** $CC(j;t,q)$ of j as the set of the residue classes of $j, j\cdot q, j\cdot q^2, \ldots$ modulo t we may write the irreducible factors of $z^t-1 \in F[z]$ as $\prod_h (z-\zeta^h) \in F(\zeta)[z]$, where h ranges over $CC(j;t,q)$ for some $j \in \mathbb{Z}_n$.
Sometimes it is convenient to define a cyclic code of block length n as the largest set of polynomials in $R_n(q)$, for which a certain set of tth roots of unity over F are zeros of a prescribed multiplicity.

2. Error Correcting Codes

2.1. Minimum Distance

Let $\mathbf{C}$ be a linear (n,k)-code over $F := GF(q)$ and let H be an $s\times n$-matrix with entries in an extension field K of F with $H\cdot\mathbf{c} = \mathbf{0}$ for every codeword $\mathbf{c} \in \mathbf{C}$. If any $\delta-1$ columns of this possibly *incomplete* check matrix H of $\mathbf{C}$ are linearly independent over F, then there is no codeword $\mathbf{c} \in \mathbf{C}\backslash\{\mathbf{0}\}$ with Hamming weight $\gamma(\mathbf{c}) < \delta$ and thus the minimum distance d of $\mathbf{C}$ is bounded from below by δ, i. e. $d \geq \delta$. If furtheron H is a complete check matrix which contains δ linearly dependent columns, then $d = \delta$.
This observation is of small value for the determination of the minimum distance of a given linear code. But it is extremely useful for the construction of cyclic **BCH-codes** with a **designed minimum distance** δ : Let n and q be relatively prime. Denote by $\mathbf{C} \subseteq R_n(q)$ the largest cyclic code for which the $\delta-1$ «consecutive» powers $\zeta^a, \zeta^{a+1}, \ldots, \zeta^{a+\delta-2}$ of the primitive n^{th} root of unity ζ over F are all roots of the code polynomials. The matrix

$$\begin{bmatrix} 1 & \zeta^{a} & \zeta^{2a} & \cdots & \zeta^{(n-1)a} \\ 1 & \zeta^{a+1} & \zeta^{2(a+1)} & & \zeta^{(n-1)(a+1)} \\ \cdot & \cdot & \cdot & & \cdot \\ \cdot & \cdot & \cdot & & \cdot \\ \cdot & \cdot & \cdot & & \cdot \\ 1 & \zeta^{a+\delta-2} & \zeta^{2(a+\delta-2)} & & \zeta^{(n-1)(a+\delta-2)} \end{bmatrix}$$

is a possibly incomplete check matrix of **C**. Any $(\delta-1)\times(\delta-1)$-submatrix can be transformed into a Vandermonde matrix, whose determinant is non-zero. Thus **C** has at least minimum distance δ. The dimension of the BCH-code **C** is the sum of the cardinalities of all cyclotomic cosets, which contain one of the exponents $a, a+1, , , a+\delta-2$. In case $a = 1$ the code **C** is called a **narrow-sense** BCH-code. In case $n = q^m-1$ for some natural number m the code **C** is called a **primitive** BCH-code. On the International Workshop on Algebraic and Combinatorial Coding Theory in Varna, Bulgaria, Sept. 1988 Ludwig Staiger made a revolutionary proposal : In honour to their inventors R. C. **B**ose **a**nd D. K. **Ray**-Chaudhuri (1960) and A. H**ocque**nghem (1959) the BCH-codes should be renamed as **BAROCQUE**-codes. "Ray" is the prefix of length three of Ray-Chaudhuri's *family*-name and in the French language the letter "H" is not pronounced. Unfortunately, Michael Kaplan made a game of this serious proposal suggesting to develop a theory of **ROKOKO**-codes. So coding theorists will have to live longer with the nearly meaningless character string "BCH".

2. 1. 1. Goppa-Codes

Let q be a power of a prime, m a natural number, $n := q^m-1$, ζ a primitive n^{th} root of unity over GF(q) and **C** the primitive narrow-sense BCH-code over GF(q) with designed distance δ. We give a necessary and sufficient condition for a polynomial $c(z) := \sum_{i=0}^{n-1} c_i z^i \in R_n(q)$ to be a codeword of **C** : Since $\sum_{i=0}^{n-1} c_i(\zeta^{k+1})^i = 0$ for $k = 0, 1, \ldots, \delta-2$ the polynomial $c(z)$ is in **C** if and only if $(z^n-1)\cdot\sum_{i=0}^{n-1} c_i/(z-\zeta^{-i}) = \sum_{i=0}^{n-1}\sum_{k=0}^{n-1} c_i(\zeta^{k+1})^i z^k \equiv 0 \bmod z^{\delta-1}$. Since $z^{\delta-1}$ and z^n-1 do not have common roots we conclude that $c(z)$ is a codeword of **C** if and only if $z^{\delta-1}$ divides the nominator of the formal rational function $\sum_{i=0}^{n-1} c_i/(z-\zeta^{-i})$ in its representation as the quotient of two polynomials. We write shortly : $c(z) \in \mathbf{C} \Leftrightarrow \sum_{i=0}^{n-1} c_i/(z-\zeta^{-i}) \equiv 0 \bmod z^{\delta-1}$.

The Goppa-codes now are a straight forward generalization of the primitive narrow-sense BCH-codes: Starting with the Galois-field $F := GF(q)$, a natural number m, a *Goppa*-polynomial $G(z) \in GF(q^m)[z]$ of degree $\deg G(z) =: s$ and a set $L = \{\gamma_0, \gamma_1, \ldots, \gamma_{n-1}\}$ of n pairwise distinct elements $\gamma_i \in GF(q^m)$ with $G(\gamma_i) \neq 0$ we define the **Goppa-code** $\Gamma(L,G(z))$ over F as the set of all vectors $\mathbf{c} = c_0 c_1 \ldots c_{n-1} \in F^n$ with $\prod_{j=0}^{n-1}(z-\gamma_j) \cdot \sum_{i=0}^{n-1} c_i/(z-\gamma_i) \equiv 0 \bmod G(z)$.
In the algebra $GF(q^m)[z]/(G(z))$ the polynomial $(z-\gamma_i)$ is invertible; its multiplicative inverse is the polynomial $(z-\gamma_i)^{-1} = -((G(z)-G(\gamma_i))/((G(\gamma_i)(z-\gamma_i))$.
Hence we may write the defining congruence as $\sum_{i=0}^{n-1} c_i/(z-\gamma_i) \equiv 0 \bmod G(z)$.
Writing down the coefficients of the polynomial entries of the vector $\mathbf{h}(z) := ((z-\gamma_0)^{-1}, (z-\gamma_1)^{-1}, \ldots, (z-\gamma_{n-1})^{-1})$ as columns we obtain an $s \times n$-matrix H over $GF(q^m)$ whose rows contain a check matrix of the Goppa-code $\Gamma(L,G)$.
We conclude $k := \dim_F \Gamma(L,G(z)) \geq n-ms$. In the same way like for BCH-codes we see that the minimum distance d of $\Gamma(L,G(z))$ is bounded from below by the value $s+1$. In case $q = 2$ and $G(z)$ separable (in particular irreducible) this bound can be sharpened: $d \geq 2s+1$.

There exists an efficient algorithm for the decoding of Goppa-codes. Assume that an error vector $\mathbf{e} = e_0 e_1 \ldots e_{n-1} \in F^n$ of Hamming weight $\gamma(\mathbf{e}) =: e \leq s/2$ was added to the codeword $\mathbf{c} = c_0 c_1 \ldots c_{n-1} \in \Gamma(L,G(z))$. The decoder then receives the vector $\mathbf{y} = y_0 y_1 \ldots y_{n-1} := \mathbf{c}+\mathbf{e} \in F^n$ and calculates in $GF(q^m)[z]/(G(z))$ its *syndrome* $S(z) := \mathbf{h}(z) \cdot \mathbf{y} = \sum_{i=0}^{n-1} y_i/(z-\gamma_i) = \sum_{i=0}^{n-1} e_i/(z-\gamma_i)$. We denote by $E := \{i; e_i \neq 0\}$ the set of positions in which the transmission errors occured. We rewrite the syndrome as $S(z) := \sum_{i \in E} e_i/(z-\gamma_i)$ and define the *error-locator* polynomial $\sigma(z) := \prod_{i \in E}(z-\gamma_i)$ of degree $\deg \sigma(z) = e$ and the *error-evaluation* polynomial $\omega(z) := \sum_{i \in E} e_i/(z-\gamma_i) \prod_{j \in E}(z-\gamma_j)$ of degree $\deg \omega(z) < e$ which are relatively prime. For all $i \in E$ we have $e_i = [\omega(z)/\frac{d}{dz}\sigma(z)]_{z=\gamma_i}$. The socalled *key-equation* says $\sigma(z) \cdot S(z) \equiv \omega(z) \bmod G(z)$. Using Euclid's algorithm for finding the greatest common divisor of $G(z)$ and $S(z)$ the decoder computes the error-locator and evaluation polynomial and reconstructs the codeword $\mathbf{c}$. For the case $q = 2$ and $G(z)$ irreducible N. Patterson developed a decoding algorithm, which allows to correct up to s errors; this algorithm is based on the facts, that the algebra $GF(q^m)[z]/(G(z))$ is a field and that the key-equation takes the form $\sigma(z) \cdot S(z) \equiv \frac{d}{dz}\sigma(z) \bmod G(z)$.

R. J. McEliece proposed a public key cryptosystem which bases on the fact, that there exist these powerful decoding algorithms for Goppa codes, but that that the best known decoding technique for general linear codes, the syndrome decoding [cf. section 1.6] is to complex to perform. The system designer chooses randomly an irreducible polynomial $G(z)$ over $GF(2^m)$ of degree s, say $m := 10$, $s := 37$, and a canonical $k \times n$ generator matrix G of the Goppa-code $\Gamma(GF(2^m), G(z))$, where $k := \dim \Gamma(GF(2^m), G(z))$ and $n := 2^m$. "Canonical" means here that the columns of G in a set K of k positions form the $k \times k$ identity matrix, the encoding map $\mathbf{Z}_2^k \ni \mathbf{m} \mapsto \mathbf{c} \in \Gamma(GF(2^m), G(z))$ places the bits of $\mathbf{m}$ unaltered in the positions of the set K in the codeword $\mathbf{c}$. Then he selects randomly a nonsingular $k \times k$ "scrambling" matrix S and an $n \times n$ permutation matrix P and publishes the matrix $G' := S \cdot G \cdot P$. The user, who wants to transmit a k-bit word $\mathbf{m}$ encrypts this message by adding an arbitrary error vector $\mathbf{e} \in \mathbf{Z}_2^n$ of Hamming weight $\gamma(\mathbf{e}) \leq s$ to the vector $\mathbf{m} \cdot G'$. The receiver recovers the original message by applying Patterson's algorithm to $\mathbf{c}' := (\mathbf{m} \cdot G' + \mathbf{e}) \cdot P^{-1} \in \Gamma(GF(2^m), G(z))$, forming with the k bits of $\mathbf{c}'$ in the positions of K the word $\mathbf{m}'$ and calculating $\mathbf{m} = \mathbf{m}' \cdot S^{-1}$.

2. 1. 2. Hamming- and Simplex-Codes

Let $r \geq 3$ be an integer. In homogeneous coordinates we write down all the $n := (q^n - 1)/(q-1)$ points of the $(r-1)$-dimensional projective space $PG(r-1,q)$ over $F := GF(q)$ as the columns of an $r \times n$-matrix H. Since any two columns of H are linearly independent and since there are three linearly dependent columns the matrix H defines as a check matrix a linear (n,k)-code over F with minimum distance $d = 3$ and dimension $k = n - r$, the socalled **Hamming code** $HAM(r,q)$. Checking the Hamming bound formula we see that that $HAM(r,q)$ is a perfect code.

After performing elementary operations on the rows of the matrix H its columns remain a system of distinct representatives of the 1-dimensional subspaces of the vector space of all r-tuples over F.

The **simplex code** $SIM(r,q)$, i. e. the dual code of $HAM(r,q)$, therefore is an equidistributed and equidistant (n,r)-code. Since all columns of H with first component zero form a hyperplane in $PG(r-1,q)$, the (minimum) distance of this optimal code has the value $d = q^{r-1}$.

2. 1. 3. MDS-Codes

We rename the elements of the Galois field $F := GF(q) := \{0,1,\alpha_3,\dots,\alpha_q\}$, choose an integer k with $2 \le k \le q-1$ and put $r := q-k+1$.
The calculation of the determinant of any $r\times r$-submatrix of the $r\times n$-matrix

$$H := \begin{bmatrix} 1 & 1 & 1 & \cdots & 1 & 0 \\ 1 & \alpha_3 & \alpha_4 & \cdots & 0 & 0 \\ \cdot & \cdot & \cdot & & \cdot & \cdot \\ \cdot & \cdot & \cdot & & \cdot & \cdot \\ \cdot & \cdot & \cdot & & \cdot & \cdot \\ 1 & \alpha_3^{r-1} & \alpha_4^{r-1} & \cdots & 0 & 1 \end{bmatrix}$$

leads to the calculation of a Van Der Monde determinant, and hence H defines as a check matrix an $(q+1,k)$-MDS-code over F.

In case $\operatorname{char} F = 2$ the $(q+2,3)$-matrix

$$H := \begin{bmatrix} 1 & 1 & 1 & \cdots & 1 & 0 & 0 \\ 1 & \alpha & \alpha & \cdots & 0 & 1 & 0 \\ 1 & \alpha_3^2 & \alpha_4^2 & \cdots & 0 & 0 & 1 \end{bmatrix}$$

defines as a check matrix a $(q+2,q-1)$-MDS-code. Since the dual code of a linear MDS-code again has the MDS property, the matrix H defines as a generator matrix a $(q+2,3)$-MDS-code.

Apart from the trivial cases $k = 0,1,n-1,n$ there are no longer k-dimensional MDS-codes over alphabets of prime power size q known.

Shorter MDS-codes can be obtained by *puncturing*, i. e. by deleting the ith component in every codeword for a fixed index i, or by *deriving*, i. e. by considering only codewords with a zero in the ith position and deleting it. Puncturing reduces the blocklength by one, deriving reduces simultaneously the blocklength an the dimension.

The problem to determine the greatest blocklength n of a linear MDS-code over $GF(q)$ of dimension $k = n-r$ is exactly B. Segre's *Problem* $I_{r,q}$:
Determine the greatest integer n *for which there exists an* n*-arc in* $PG(r,q)$.

The results obtained until now support the conjecture $n = q+2$ for $\operatorname{char} F = 2$ and $k = 3$ or $k = q-1$ and $n = q+1$ for $\operatorname{char} F \neq 2$.

From a practical point of view the cyclic MDS-codes are extremely interesting. Let n be a divisor of $q-1$. Then F contains all the nth roots of unity and for every r with $0 \le r \le n$ the BCH-code with generator polynomial $\prod_{h=1}^{r}(z-\zeta^h)$ has the designed minimum distance $\delta = r+1$ and hence is MDS.

In case $n = q-1$ these cyclic codes are the socalled **Reed- Solomon-codes**. For the error correction in Compact Disc systems two linear (32,28)- resp. (28,24)-codes which are derived from the (255,251)-Reed-Solomon-code over $GF(2^8)$ are in use.

If n is a divisor of $q+1$, then the cyclotomic coset of $j \in \mathbb{Z}_n$ has the form $CC(j;n,q) = \{j,-j\}$. Hence for any odd integer $r \le n$ the polynomial

$$\prod_{h=-(r-1)/2}^{(r-1)/2} (z-\zeta^h)$$

generates in $R_n(q)$ an MDS-code.

The same applies for r even, $0 \le r < n$ and n odd to the polynomial

$$\prod_{h=(n-r+1)/2}^{(n+r-1)/2} (z-\zeta^h).$$

A deep number theoretic result of E. F. Assmus, Jr. and H. F. Mattson, Jr. says that *for any prefixed prime number* n *there exists a integer* p_0, *such that for any prime number* $p \ge p_0$ *all cyclic* (n,k)*-codes over a finite field of characteristic* p *have the MDS property*. Unfortunately this number p_0 can be very large. E. g. there is a cyclic (503,252)-code over the Galois field GF(43030081169) which is not MDS. On the other hand, communication engineers are interested in codes over small fields, but a simple combinatorial argumentation yields the inequality $n \le q+k-1$ for every non trivial (n,k)-MDS-code over an alphabet of size q.

2. 1. 4. Reed-Muller-Codes

Let $\mathbf{C}_1, \mathbf{C}_2$ be two linear codes of dimension k_1 resp. k_2, minimum distance d_1 resp. d_2 and the same blocklength n over $F := GF(2)$. We define the **sum construction** of $\mathbf{C}_1$ and $\mathbf{C}_2$ as the binary $(2n,k_1+k_2)$-code

$$\mathbf{C}_1 \& \mathbf{C}_2 := \{(\mathbf{x}_1, \mathbf{x}_1+\mathbf{x}_2) ; \mathbf{x}_1 \in \mathbf{C}_1, \mathbf{x}_2 \in \mathbf{C}_2\}.$$

We observe that in $\mathbf{C}_1 \& \mathbf{C}_2$ a codeword of minimum weight must be of type $(\mathbf{x}_1, \mathbf{x}_1)$ or $(\mathbf{0}, \mathbf{x}_2)$ and hence the sum construction of $\mathbf{C}_1$ and $\mathbf{C}_2$ has minimum distance $d(\mathbf{C}_1 \& \mathbf{C}_2) = \min\{2d_1, d_1+d_2\}$. Using this method of combining two codes, we recursively define for $0 \le s \le m$ the **Reed Muller code** $RM(m,s)$ of *order* s and length $n := 2^m$ over F: As initial values we set

$$RM(m,0) := \{\mathbf{0},\mathbf{1}\} \subseteq F^n \;,\; RM(m,m) := F^n \quad \text{for} \quad m \ge 0.$$

Then we define

$$RM(m+1,s+1) := RM(m,s+1) \& RM(m,s) \quad \text{for} \quad 0 \le s \le m.$$

We observe the analogy of this definition with the standard recursion relation of the binomial coefficients and get the minimum distance $d(RM(m,s)) = 2^{m-s}$ and the dimension $k = \sum_{i=0}^{s}\binom{m}{i}$.

Geometrically the Reed-Muller-Code $RM(m,s)$ can be described as the set of incidence vectors of all at least $(m-1)$-dimensional affine subspaces of the m-dimensional affine space $AG_m(2)$ over F and their symmetric differences. I. S. Reed developed a powerful decoding algorithm based on majority vote decisions. Because of its easy implementation the NASA used the Reed Muller code $RM(5,1)$ for some deep space missions, although the error correcting capabilities of this code are far from being optimal.

2. 2. Optimal Geometries

Any code $\mathbf{C} \subseteq F^n$ can be interpreted as a geometrical structure $\mathfrak{G}(\mathbf{C})$. The set of **points** of this geometry is the cartesian product $P := F\times\mathbb{Z}_n$. Its set of **blocks** is the set $\mathbf{C}$ of the codewords. For a point $(\alpha,i) \in P$ and a block $\mathbf{c} = c_0c_1 \ldots c_{n-2}c_{n-1} \in \mathbf{C}$ we define the **incidence** by $(\alpha,i) \in \mathbf{c} :\Leftrightarrow c_i = \alpha$. In other words : The structure $\mathfrak{G}(\mathbf{C})$ is the geometry of the graphs of the mappings $\mathbf{c} : \mathbb{Z}_n \to F\,;\, i \mapsto c_i$.

The geometry $\mathfrak{G}(\mathbf{C})$ is a generalization of the socalled *group divisible* or *transversal designs*: No block intersects for any $i \in \mathbb{Z}_n$ the *generating line* $F_i := \{(\alpha,i)\,;\,\alpha \in F\}$ in more than one point.

One cannot expect miraculous results from this description; it is a mere translation of the coding theoretical language into the geometrical terminology. But in the case of an (n,k)-MDS-code $\mathbf{C}$ over an alphabet F of size q the geometry $\mathfrak{G}(\mathbf{C})$ has been studied intensively under the name of an **optimal geometry**. In fact, the separability condition says, that for any k points, no two of them belonging to the same generating line, there is precisely one block incident with them all. For $k = 2$ and $n = q$ or $n = q+1$ the optimal geometry $\mathfrak{G}(\mathbf{C})$ is an affine plane of order q with one parallel pencil of lines serving as generating lines or a punctured projective plane of order q with the pencil of lines passing through the missing point serves as the set of generating lines, respectively. In general, giving an $(n,2)$-MDS-code over F is the same like giving an n-web (n-*Gewebe*, n-*rete*) or $n-2$ mutually orthogonal latin squares of order q. For $k=3$ and $n=q+1$ the geometry $\mathfrak{G}(\mathbf{C})$ is a Laguerre plane of order q. More general : For an arbitrary $k \geq 2$ and $n=q+k-2$ the geometry $\mathfrak{G}(\mathbf{C})$ is a Laguerre-(k-2)-structure of order q. The numerous research results on the existence and structure of the optimal geometries can directly be retranslated into the language of coding theory.

2. 3. Automorphisms

As a rule of experience the implementation of a code into a communication system will be the more facilitated the more homogeneous it is, i. e. if it has an automorphism group with a high degree of transitivity. Here by a **code isomorphism** $\varphi : \mathbf{C} \to \mathbf{D}$ between two linear codes $\mathbf{C}, \mathbf{D} \in F^n$ we mean a linear isometry with respect to the Hamming metric. As an analogue to Witt's theorem on isometries in metric vectorspaces *any code isomorphism* $\varphi : \mathbf{C} \to \mathbf{D}$ *can be extended to an isometry* Φ *of the whole space* F^n. Thus φ is a **monomial** transformation, i. e. it can be represented by the product $\Pi \cdot \Delta$ of an $n \times n$ permutation matrix Π with an $n \times n$ diagonal matrix Δ. In other words : there exists a permutation $\pi \in S_n$ and n elements $\delta_0, \delta_1, \ldots, \delta_{n-1} \in F \backslash \{0\}$ with $\varphi(c_0 c_1 \ldots c_{n-1}) = (\delta_0 c_{\pi(0)}, \delta_1 c_{\pi(1)}, \ldots, \delta_{n-1} c_{\pi(n-1)})$ for all codewords $c_0 c_1 \ldots c_{n-1} \in \mathbf{C}$. The geometries $\mathfrak{G}(\mathbf{C})$ and $\mathfrak{G}(\mathbf{D})$ differ by the arrangement of the generating lines which themselves are permuted each by a dilatation. For a given isomorphism $\varphi : \mathbf{C} \to \mathbf{D}$ the matrices Π and Δ are uniquely determined, if the minimum distance $d^{\perp}$ of the dual code $\mathbf{C}^{\perp}$ is greater than two. In any case, the monomial map $\Pi^{-1} \cdot \Delta$ induces a code isomorphism $\mathbf{C}^{\perp} \to \mathbf{D}^{\perp}$. Hence, if $d, d^{\perp} \geq 3$ then for every linear code $\mathbf{C}$ the groups $\mathrm{Aut}(\mathbf{C})$ and $\mathrm{Aut}(\mathbf{C}^{\perp})$ are isomorphic. Since the simplex code is equidistant we obtain as an application $\mathrm{Aut}(\mathrm{SIM}(r,q)) \cong \mathrm{Aut}(\mathrm{HAM}(r,q) \cong GL_r(q)$.

The vector space over $GF(q)$ which is generated by the incidence vectors of the blocks of a geometrical design often turns out to be a linear code with reasonable error correcting capabilities and good implementation possibilities. A standard example is the above mentioned Reed Muller code RM(m,s). Using its geometrical description one determines its automorphism group :

$$\mathrm{Aut}(\mathrm{RM}(m,s) \cong AGL_m(2) \quad \text{for} \quad 0 < s < m-1 .$$

Although the Reed-Muller-Codes are one of the best studied class of codes, the distribution of the weights of the codewords of the third order Reed-Muller-Codes is still unknown.

Cyclic codes are defined as linear codes for which the cyclic shift

$$F^n \to F^n\ ;\ x_0x_1\ldots x_{n-2}x_{n-1} \mapsto x_{n-1}x_0x_1\ldots x_{n-2}$$

is an automorphism. The *equivalence theorem* tells us under which conditions two cyclic codes in $R_n(q)$ are isomorphic :

Let v *be the greatest common divisor of* n *and* q, $t := n/v$ *and let* ζ *and* η *be two primitive roots of unity over* $GF(q)$ *and* $\mathbf{C}_\zeta, \mathbf{C}_\eta$ *two cyclic codes in* $R_n(q)$. *If for every* $h = 0,1,\ldots,t-1$ *the element* ζ^h *is a root of* $\mathbf{C}_\zeta$ *of the same multiplicity as* η^h *of* $\mathbf{C}_\eta$, *then* $\mathbf{C}_\zeta \cong \mathbf{C}_\eta$.

3. Quadratic residue codes

The class of the *quadratic residue codes* (QR-codes) is one of the most promising series of cyclic codes. With an information rate $\sim 1/2$ they are defined over a prime field $GF(p)$ or a quadratic extension $GF(p^2)$ thereof and have a prime number length n, hence by the Asmuss-Mattson-theorem are suspicious to be MDS, if the characteristic p exceeds a certain limit.

3. 1. Quadratic Residues

Let $n = 4t + (\frac{-1}{n})t \geq 3$ be a prime number, Q the multiplicative subgroup of quadratic residues in the field $\mathbb{Z}_n$ of the residue classes of integers modulo n and $N := \mathbb{Z}_n \setminus (Q \cup \{0\})$ its coset, consisting of the nonresidues modulo n. Furtheron let p be a prime number distinct from n; for the sake of clarity we denote by $GF(p)$ the prime field of characteristic p. If m is the least positive integer such that $p^m \equiv 1 \bmod n$, then the splitting field $GF(p^m)$ of the polynomial $z^n - 1 \in GF(p)[z]$ over $GF(p)$ is the nth cyclotomic field of characteristic p. We denote by $\zeta \neq 1$ an nth root of unity over $GF(p)$.

We can exchange the rôles of the **Gaussian periods**

$$\xi := \sum_{i\in Q} \zeta^i \ , \quad \xi' := \sum_{i\in N} \zeta^i$$

by simply substituting ζ by $\eta := \zeta^u$ with $u \in N$.

We have $\xi+\xi' = -1$ and $\xi\cdot\xi' = \sum_{(i,j)\in Q\times N} \zeta^{i+j} = -(\frac{-1}{n})(n-1)/4$. Thus ξ and ξ' are the roots of the polynomial $z^2+z-(\frac{-1}{n})t \in GF(p)[z]$ and the identity $(2\xi+1)^2 = (\frac{-1}{n})n$ follows. We always have $\xi \neq \xi'$. Since

$$\xi^p = \begin{cases} \xi & \text{if} \quad p \in Q \\ \xi' & \text{if} \quad p \in N \end{cases}$$

we see that ξ belongs to $GF(p)$, i. e. that the polynomial $z^2+z-(\frac{-1}{n})t$ is reducible over $GF(p)$, if and only if $p \in Q$, i. e. $\xi \in GF(p) \Leftrightarrow (\frac{p}{n}) = 1$. This is an algebraic formulation of the Gaussian quadratic reciprocity law together with its supplement for $p = 2$.

We set $q := \begin{cases} p & \text{if} \quad p \in Q \\ p^2 & \text{if} \quad p \in N \end{cases}$ and $F := GF(q) = GF(p)(\xi)$.

3. 2. QR-Codes

With the previous notations, we set $r := (n-1)/2$, $k := r+1 = (n+1)/2$ and define the **quadratic residue code** Q of length n over F as the cyclic (n,k)-code in $R_n(q)$ with the generator polynomial

$$g(z) := \prod_{h\in Q} (z-\zeta^h).$$

The polynomial $(z-1)g(z)$ generates the **expurgated QR-code** Q^- The cyclic code N generated in $R_n(q)$ by $g_N(z) := (z^n-1)/((z-1)g(z))$ is isomorpic to Q. We define $N^- := \{c(z) \in N \, ; \, c(1) = 0\}$. Inspecting the isometric algebra automorphism $R_n(q) \to R_n(q) \, ; \, x(z) \mapsto x(z^{-1})$ we find

$$Q^\perp \cong \begin{cases} Q^- & \text{if} \quad n \equiv -1 \bmod 4 \\ N^- & \text{if} \quad n \equiv 1 \bmod 4 \end{cases}$$

We can write the **extended QR-code** as

$$\hat{Q} := \{c_0c_1...c_{n-1}c_\infty ; c_0c_1...c_{n-1} \in Q , c_\infty = (2\xi+1)\sum_{i=0}^{n-1} c_i/n\}$$

or shortly as $\hat{Q} = \{(c(z), (2\xi+1)\cdot c(1)/n) ; c(z) \in Q\}$. This definition assures that the $(n+1,k)$-code $\hat{Q}$ is self dual for $n \equiv -1 \bmod 4$ while for $n \equiv 1 \bmod 4$ it is orthogonal to $\check{N} := \{(c(z), -(2\xi+1)\cdot c(1)/n) ; c(z) \in N\}$.
The isomorphisms $Q \to N$ induce in a natural way isomorphisms $\hat{Q} \to \check{N}$.

3. 3. The Square Word

With the previous notations we define the **square word**

$$q(z) := \xi + \sum_{h \in Q} z^h$$

and the **non square word**

$$n(z) := -\xi' + \sum_{h \in N} z^h$$

For $u \in \mathbb{Z}_n \setminus \{0\}$ the element ζ^u is a root of $q(z)$ resp. $n(z)$ if and only if $u \in Q$. Both $q(z)$ and $n(z)$ are thus codewords of Q. Further we have $q(1) = 0$ resp. $n(1) = 0$ if and only if $\xi = (n-1)/2$ resp. $\xi' = (n-1)/2$.

$$\text{It is} \quad \{\xi,\xi'\} = \{(n-1)/2, -(n-1)/2\} \iff \begin{cases} p = 2 \quad \text{and} \quad n \equiv \pm 1 \bmod 8 \\ \text{or} \\ p > 2 \quad \text{and} \quad n \equiv -\left(\frac{-1}{n}\right) \bmod p \end{cases} .$$

In case of equality we can replace the nth root of unity ζ by ζ^u with $u \in N$ and thus determine which one of the relations $q(1) \neq 0$, $n(1) \neq 0$ is to hold. If the choice of ζ is such that $q(1) \neq 0$ holds, then the square word $q(z)$ generates in $R_n(q)$ the QR-code Q as an ideal, i. e.

$$Q = \{t(z)\cdot q(z) ; t(z) \in R_n(q)\};$$

in other words : The n cyclic shifts $q_h(z) := z^h \cdot q(z)$, $h \in \mathbb{Z}_n$, of the square word $q(z)$ form a set of generators for the k-dimensional subspace of the vector space $R_n(q)$. If instead ζ is chosen in such a way that $\xi = (n-1)/2$, then the vectors $q_h(z)$, $h \in \mathbb{Z}_n$, generate the r-dimensional expurgated QR-code Q^-. In any case together with the the *all ones word* $(z^n-1)/(z-1)$ these cyclic shifts always generate the QR-code Q as a vector space.

The square word $q(z)$ satisfies the identity $q(z)^q = q(z)$. It follows, that $q(z)$ generates a cyclic subgroup of the multiplicative semigroup of the cyclic code Q or Q^- generated by $q(z)$ in $R_n(q)$. Thus $q(z)^{q-1}$ coincides with the identity $i(z)$ of this semigroup, the idempotent of the code. Clearly, for $q = 2$ these polynomials coincide; but for $q > 2$ it is much easier to work with the square word than with the idempotent, since the coefficients of $i(z)$ are hard to determine.

The additional check symbol of the *extended square word* $\hat{q} := (q(z), q_\infty) \in \hat{Q}$ turns out to be
$$q_\infty = \begin{cases} \xi & \text{for } n \equiv 1 \mod 4 \\ 1+\xi & \text{for } n \equiv -1 \mod 4 \end{cases}.$$

In case $\xi \neq (n-1)/2$ the vectors $\hat{q}_h := (q_h(z), q_\infty)$, $h \in \mathbb{Z}_n$, generate the vector space $\hat{Q}$. The additional check symbol of the all one word $(z^n-1)/(z-1) \in Q$ in $\hat{Q}$ is $1_\infty := 2\xi+1$, while in $\check{N}$ the all one word is extended by the symbol $-2\xi-1$.

3. 4. The Gleason-Prange-Theorem

Obviously all the permutations $\mathbb{Z}_n \to \mathbb{Z}_n$; $i \mapsto a \cdot i + b$, $a \in Q$, $b \in \mathbb{Z}_n$, induce automorphisms of the QR-code Q. The Gleason-Prange-Theorem states, that *the the automorphism group of the extended QR-code* $\hat{Q}$ *has a subgroup, for which the group of the permutation parts of its (monomial) mappings is isomorphic to the special linear group* $PSL_2(n)$.

The crucial step in the proof is to show, that the monomial transformation
$$\varphi : (x_0, x_1, \ldots, x_i, \ldots, x_{n-1}, x_\infty) \mapsto (-x_\infty, x_{-1}, \ldots, (\tfrac{i}{n}) x_{-1/i}, \ldots, (\tfrac{n-1}{n}) x_{-1/(n-1)}, -(\tfrac{-1}{n}) x_0)$$
maps the code $\hat{Q}$ onto itself. We assume $\xi \neq (n-1)/2$ and show $\varphi(\hat{q}_h) \in \hat{Q}$ for all $h \in \mathbb{Z}_n$:

Indeed, we have : $$\varphi(\hat{q}_0) = \begin{cases} \hat{q}_0 & \text{if } n \equiv 1 \mod 4 \\ \hat{q}_0 - ((z^n-1)/(z-1), 1_\infty) & \text{if } n \equiv -1 \mod 4 \end{cases}$$

and for $h \in \mathbb{Z}_n \setminus \{0\}$: $$\varphi(\hat{q}_h) = \begin{cases} \hat{q}_0 + \hat{q}_{-1/h} - ((z^n-1)/(z-1), 1_\infty) & \text{if } -h \in Q \\ \hat{q}_0 + \hat{q}_{-1/h} & \text{if } -h \in N \end{cases}$$

The importance of the Gleason-Prange-Theorem lies in the fact that the automorphism group of the extended QR-code acts 2-transitively on the set of positions $\mathbb{Z}_n \cup \{\infty\}$.
The binary QR-code of length 7 is a cyclic version of HAM(3,2). The full automorphism group of the extended code is isomorphic to $GL_3(2)$. The full automorphism groups of the extended **Golay-Codes**, i. e. the extended binary resp. ternary QR-codes of length 24 resp. 12 are the Mathieu groups M_{24} resp. M_{12}. The alternate group of degree 6 acts as an automorphism group on the positions of the extended QR-code of length 6 over GF(4). These are the four known cases in which the Gleason-Prange-Theorem does not describe the full automorphism group of the extended QR-code.

3.5. Minimum Distance

Evidently, the minimum distance $\hat{d} := d(\hat{Q})$ of the extended QR-code $\hat{Q}$ exceeds the minimum distance d of the original QR-code Q at most by the value 1.
Assume that there exists a codeword $\hat{\mathbf{c}} = c_0 c_1 \ldots c_{n-1} c_\infty \in \hat{Q}$ of Hamming weight $\gamma(\hat{\mathbf{c}}) = d$. Since Q is the puncturing of $\hat{Q}$ in the position ∞, we conclude $c_\infty = 0$. By the Gleason-Prange-Theorem $\mathrm{Aut}(\hat{Q})$ acts transitively on the set of positions $\mathbb{Z}_n \cup \{\infty\}$. We cancel a non zero component in $\hat{\mathbf{c}} = c_0 c_1 \ldots c_{n-1} c_\infty = c_0 c_1 \ldots c_{n-1} 0$ and contradictionarily we obtain a codeword of weight $d-1$ in a puncturing of $\hat{Q}$, which is isomorphic to the QR-code Q. Hence $\hat{d} = d+1$. Similarily $d^- := d(Q^-) = d+1$.

In order to calculate the minimum distance of a QR-code Q of length n over $F := GF(q)$ we restrict ourselves to the case $n \equiv -1 \bmod 4$, since here we have $\hat{Q}^\perp = \hat{Q}$ and the theory of these QR-codes is further developed than in the case $n \equiv 1 \bmod 4$.

Let $\mathbf{c} = c_0 c_1 \ldots c_{n-1} \in Q$ be a codeword of minimum weight $\gamma(\mathbf{c}) = d$. The residue classes of $\mathbb{Z}_n$ are called **points**, the d points in the **support** $V := \mathrm{supp}(\mathbf{c}) := \{i \in \mathbb{Z}_n ; c_i \neq 0\}$ of $\mathbf{c}$ are called **vertices**. The n supports $V+h$, $h \in \mathbb{Z}_n$, of the cyclically shifted codeword $\mathbf{c}$ are called **blocks**. The map $V \cap (V+h) \to V \cap (V-h)$; $i \mapsto i-h$ is always bijective. If for $h \neq 0$ the block $V+h$ contains exactly one vertex i, $V \cap (V+h) = \{i\}$, then $j := i-h$ is a vertex, too, we have $V \cap (V-h) = \{j\}$ and call the set $\{i,j\}$ an **edge**. The QR-code $\hat{Q}$ is selfdual. Therefore for any edge $\{i,j\}$ we have $c_i \cdot c_j = -c_\infty^2$. The vertices and the edges form the **minimal graph**. The minimal graph is replaced by an isomorphic graph, if we substitute the codeword **c** by a cyclic shift. Thus we are allowed to choose an arbitrary position $h \in \mathbb{Z}_n$ and to assume $c_h = 1$.

The minimal graph possesses $e := (n-1-\alpha)/2$ edges, where α is the number of blocks distinct from V, which intersect V in at least two vertices.

In order to get a lower bound for e we determine an upper bound for α: A vertex $i \in V$ is incident with the block $V+h$ if and only if $i-h$ is again a vertex. Thus every of the d vertices is incident with exactly $d-1$ blocks distinct from V and hence $\sum_{h=1}^{n-1} |V \cap (V+h)| = d \cdot (d-1)$. We conclude

$$\alpha \leq \sum_{h=1}^{n-1} (|V \cap (V+h)|-1) = d \cdot (d-1) - n + 1.$$

Since $\alpha \geq 0$ we get as a side result the **square root bound**

$$n \geq d \cdot (d-1) + 1$$

The desired lower bound for the number of edges is

$$n-1-\binom{d}{2} \leq e.$$

The only QR-codes of length $n \equiv -1 \bmod 4$ for which the square root bound cannot be improved are the trivial (3,2)-QR-codes and the Hamming code HAM(3,2) of length 7 over GF(2).

For a hypothetical QR-code Q with $n = d \cdot (d-1) + 1$ the points and blocks form a projective plane of order $d-1$. The Bruck-Ryser-theorem on the non-existence of projective planes then sometimes yields the non-existence of Q, e. g. there is no QR-code of length 13111 with minimum distance 115. This argumentation does not make any use of the underlying field $F = GF(q)$.

By a theorem of Túrán the minimal graph of a QR-code Q contains a triangle $\Delta = \{h,i,j\}$ if $e > \lfloor d^2/4 \rfloor$ and this is surely the case if $3d^2 - 2d < 4(n-1)$. Then $c_h c_i = c_i c_j = c_j c_h = -c_\infty^2$ and hence $c_h = c_i = c_j$. We may assume $c_h = 1$ and conclude that $-1 = c_\infty^2$ is a square in F. If F is the prime field GF(p) then this yields $p \equiv 1 \bmod 4$. Altogether we have the following improvement of the square root bound:

The minimal distance d *of any QR-code of length* $n \equiv -1 \bmod 4$ *over a prime field* GF(p) *with* $p \equiv -1 \bmod 4$ *satisfies the inequality* $d(3d-2) \geq 4(n-1)$.

Now let Q be a quadratic residue code of length n over GF(q) with the least minimum distance $d = \lceil (1+\sqrt{4n-3})/2 \rceil$ possible by the square root bound. If $n \neq 11$ and $n > d(d-2)$ then the minimal graph of Q contains a triangle and is connected. Thus we can assume that all non zero components of **c** have the value 1. From this we can conclude

$q = p$, $d \equiv 1 \bmod n$, $(d(d-1)+1-n)/2 \equiv 0 \bmod n$, a set of conditions, which in the range $7 \leq n \leq 379$ is only possible for the triples $(n,d,q) = (7,3,2), (31,6,5), (103,11,2), (199,15,2), (307,18,17)$. There are other results which lead to the conjecture, that for $n \geq 7$ the binary Hamming-code HAM(3,2) is the only QR-code of block length $n \equiv -1 \bmod 4$, for which the square root bound is sharp.

Computer aided investigations have been made to determine the exact value of the minimum distance d. Up to the value $n = 31$ it has been calculated for all characteristics. Other calculations using rather complicated formulas for the coefficients of the generator polynomial show that there exist QR-codes of very high lengths over relatively small fields, but nevertheless do not enjoy the MDS property, e. g. $n = 43\,030\,081\,169$ and $q = 503$.

3. 6. The Golay Codes

The **binary Golay code** GOL(23) is defined as the QR-code of length 23 over GF(2).

In order to calculate its generator polynomial g(z) we factorize $z^{23}-1$ over GF(2) : The cyclotomic cosets in $\mathbb{Z}_{23}$ are the sets {0}, the group Q of the quadratic residues modulo 23 and its coset N. The check polynomial h(z) is the reversion of the generator polynomial $g(z) = \sum_{i=0}^{11} g_i z^i$. Since we have $g(z) \cdot h(z) = (z^n-1)/(z-1)$ we obtain the relations $\sum_{i=0}^{h} g_h \cdot g_{11-h+i} = 1$ for $h = 0,1,\ldots$ We choose the 23rd root of unity ζ in such a way that the Gaussian sum ξ takes the value 0. With all these informations we get $g(z) = 1+z+z^5+z^6+z^7+z^9+z^{11}$. The square word has the form

$$q(z) = \sum_{h \in Q} z^h = z+z^2+z^3+z^4+z^6+z^8+z^9+z^{12}+z^{13}+z^{16}+z^{18} = (z+z^3+z^7) \cdot g(z).$$

The code GOL(23) is a BCH-code of designed minimum distance $\delta = 5$. By the square root bound we get $d \geq 6$. Now $d = 6$ is impossible since for any binary QR-code Q of length $n \equiv \pm 1 \bmod 4$ every codeword $\mathbf{c} \in Q \backslash Q^-$ has odd Hamming weight : Since in GF(2) we have $2\xi+1 = 0$ the additional check bit in the position ∞ of the all ones codeword $(z^n-1)/(z-1) \in Q \cap N$ has the value 1, in $\hat{Q}$ as well as in $\check{N}$. Thus $\hat{\mathbf{c}}$ is orthogonal to the all ones word of length $n+1$ and has even Hamming weight $\gamma(\hat{\mathbf{c}})$. Since $\mathbf{c} \notin Q^-$ we have $c_\infty \neq 0$, whence $\gamma(\mathbf{c}) = \gamma(\hat{\mathbf{c}})-1$. In particular the minimum distance d of GOL(23) must be odd. The generator polynomial $g(z)$ has weight 7, therefore $d = 7$. The parameters $q = 2$, $n = 23$, $d = 7$ fulfill the Hamming bound with equality. Thus the binary Golay code GOL(23) is a perfect 3-error-correcting code. Its automorphism group is the Mathieu group M_{23}.

We denote by $\mathbf{x}^*$ the word, which emerges from $\mathbf{x} \in \mathbb{Z}_2^4$ by exchanging the bits 0 and 1. The linear code

$$\mathbf{C}_1 := \{(\mathbf{x},\mathbf{y}) \in \mathbb{Z}_2^8 \,;\, \mathbf{x} \in \mathbb{Z}_2^4,\ \mathbf{y} = \mathbf{x},\mathbf{x}^* \text{ if } \gamma(\mathbf{x}) \equiv 0,1 \bmod 2\}$$

then is an equivalent version of the extended binary (8,4)-Hamming-code. We combine the code $\mathbf{C}_1$ with the equivalent extended Hamming code

$$\mathbf{C}_2 := \{x_0x_1x_2x_3y_0y_1y_2y_3 \,;\, x_0x_1x_2x_3y_3y_0y_1y_2 \in \mathbf{C}_1\}$$

to an equivalent version

$$\{(\mathbf{c}_1+\mathbf{c}_2, \mathbf{d}_1+\mathbf{c}_2, \mathbf{c}_1+\mathbf{d}_1+\mathbf{c}_2) \,;\, \mathbf{c}_1, \mathbf{d}_1 \in \mathbf{C}_1, \mathbf{c}_2 \in \mathbf{C}_2\}$$

of the extended binary Golay code. This description suggests simple coding and decoding algorithms.

The **ternary Golay code** GOL(11) is defined as the QR-code of length 11 over GF(3). Since in its extension every codeword is isotrop, its minimum distance d satisfies the congruence $d \equiv -1 \bmod 3$. The BCH bound and the square root bound both yield $d \geq 4$. The Hamming bound yields $d \leq 5$. Hence $d = 5$. Thus GOL(11) is a perfect code, too.

References

Assmus, E. F. Jr., Mattson, H. F. Jr. : *New 5-Designs.* J. Comb. Theory **6** (1969) 122-151.

Halder, H.-R., Heise, W. : *Einführung in die Kombinatorik.* München, Wien : Hanser 1976 and Berlin : Akademie 1977.

Heise, W., Quattrocchi, P. : *Informations- und Codierungstheorie.* 2. Aufl., Berlin, Heidelberg, New York : Springer 1989.

Heise, W., Zehendner, : *Quadratsummen in GF(p).* Mitt. Math. Sem. Giessen **164** (1984) 185-189.

Heise, W., Kellerer, H. : *Eine Verschärfung der Quadratwurzel-Schranke für Quadratische-Rest-Codes einer Länge* $n \equiv -1 \bmod 4$. J. Inf. Process. Cybern. EIK **23** (1987) 113-124.

Heise, W., Kellerer, H. : *Über die Quadratwurzel-Schranke für Quadratische-Rest-Codes.* J. of Geometry **31** (1988) 96-99.

Kaplan, M. : *A Class of Cosets of Second Order Reed-Muller-Codes.* To appear in Atti Sem. Mat. Fis. Univ. Modena.

Klemm, M. : *Über die Wurzelschranke für das Minimalgewicht von Codes.* J. Comb. Theory **A 36** (1984) 364-372.

MacWilliams, F. J. , Sloane, N. J. A. : *The Theory of Error-Correcting Codes.* Amsterdam, New York, Oxford : North Holland 1977.

McEliece, R. J. : *The Theory of Information and Coding.* London, Don Mills, Sydney, Tokyo : Addison-Wesley 1977.

McEliece, R. J. : *A Public-Key Cryptosystem Based on Algebraic Coding Theory.* DSN Progress Report **42-44** (1978) 114-116

Newhart, D. W. : *On Minimum Weight Codewords in QR Codes.* J. Comb. Theory **A 48** (1988) 104-119

Staiger, L. : *On the Square-Root Bound for QR-Codes.* J. of Geometry **31** (1988) 172-178.

AN INTRODUCTION TO ARRAY ERROR CONTROL CODES

P. G. Farrell
University of Manchester, Manchester, U.K.

ABSTRACT

Array error control codes are linear block and convolutional codes which are constructed from several single parity check or other component codes, assembled in two or more geometrical dimensions or directions, with emphasis on simple component codes and low complexity methods of decoding. Array codes were first discussed by Elias [31], and useful introductory references include [15,36,38,75,81]. The following books have significant sections devoted to array (or product) codes: Pohlmann [67], Rao and Fujiwara [69] and Watkinson [84]. Array codes are easy and flexible to design, and relatively simple (and therefore fast) to decode. For these reasons they have been used in a number of applications where these properties are highly desirable, particularly where information symbols appear (or are constrained to be) in geometrical patterns (e.g., tapes, cards, discs, chips, etc.). The two main fields of application are in communications systems and for information storage systems.

1. SINGLE PARITY CHECKING IN SEVERAL DIRECTIONS

1.1 The Single-Parity-Check Code

A single binary parity check in one dimension, as in Figure 1, forms an $(n, n-1, 2) = (k+1, k, 2)$ single-parity-check (SPC) code, sometimes called a simple parity code. The parity may be even or odd; even parity makes the code linear and is more convenient mathematically, but odd parity is sometimes used to improve the synchronisation and spectral properties of the code. Even parity will be assumed in what follows below. In Figure 1, the parity bit is placed at the end of the block, so that the codeword is in systematic form. The parity bit may be placed in any convenient position in the word, however; the only essential requirement is that the word have even weight. The SPC can detect any odd number of errors in a codeword, or it can correct a single erasure.

1.2 The Row-and-Column Code

An obvious example of combined SPC codes is single parity checking in two dimensions, as in Figure 2, which forms a row-and-column parity array code, also called a two-coordinate, bidirectional, bit and block parity, or crossword code [36,54,59,67,74,85]. The code may be square or rectangular, and has parameters $(n_1n_2, (n_1-1)(n_2-1), 4) = ((k_1+1)(k_2+1), k_1k_2, 4)$. This code is the simplest example of a product code [2,3,12,14,20,31,52,79]. Thus the parameters of the row-and-column (RAC) array code are the product of the corresponding parameters of the component SPC codes, and its generator matrix is the Kroenecker product of the SPC generator matrices (Slepian, 1960). In particular, the distance of the RAC code is $d = 2x2 = 4$, which is confirmed by its ability to correct a single error (the corresponding row and column parities fail) and detect double errors, or detect up to three errors (at least one row and one column parity fails). The RAC code also detects all other error patterns except those which are codewords in the code (i.e., have even weight in both dimensions). Alternatively, it can correct up to three random erasures. For a given value of $n = n_1n_2$, the rate $R = k/n = k_1k_2/n$ of the code is maximised when $k_1 = k_2 = k_0$ and $n_1 = n_2 = n_0$ (i.e., a square code [15]. Removal of the check-on-checks bit (see Figure 2) reduces the distance of the code from 4 to 3. Single error correction (SEC) is still possible, essentially because the row and column checks form a set of two orthogonal parity check equations on each information bit in the array. Two orthogonal parity checks permits single error correction by means of the majority logic (threshold) decoding technique (both parities associated with a particular information bit must fail if the bit is in error), just as J orthogonal parity checks permits correction of $\lfloor J/2 \rfloor$ random errors (an information bit is corrected if more than J/2 of the parity checks involving it fail [52,55].

1.3 Array Codes with Orthogonal Parity Checks

The previous remark indicates a way of increasing the error-correcting power of the RAC code: form additional orthogonal single parity checks on the array of information bits, by taking sufficient diagonal directions (dimensions). Figure 3 shows an array code based on a 5x5 information bit array, with 6 sets of orthogonal parity checks: the row and column checks, the two main diagonal checks, and two of the "knights move" checks (only a few are drawn, for clarity). The code is capable of correcting 3 random errors, or 6 random erasures, and has parameters $n = 25 + 10 + 18 + 26 = 79$, $k = 25$, $d = 7$. For a $k_1 x k_2$ information bit array, there are 2 sets of $k_1+k_2 - 1$ main ("one-one") diagonals, 4 sets of $k_1+2(k_2-1)$ "knights move" ("one-two") diagonals, 4 sets of $k_1+3(k_2-1)$ "one-three" diagonals, etc. This is clearly not an efficient way to proceed, for at least two closely related reasons: diagonal parities on the corners of the array check only one (or a few) information bits; and the redundancy of the code is increasing much more rapidly than the corresponding increase in error correction capability. One answer to the first problem is to "round off" an array, by removing the inefficiently checked information bits on the corners, which corresponds to shortening and puncturing [3] the original array code. Figure 4 shows the (74, 24, 7) code derived in this way from the 6x6 information bit array code with parameters (102, 36, 7). The (74, 24, 7) array code with rate 0.324 saves 14 parity checks, for one less information bit, compared to the (79, 25, 7) code with rate 0.287 based on the complete 5x5 information array. As J increases, the optimal shape for array codes of this type approaches that of a 2J-sided regular polyhedron; clearly this requires very large information bit arrays [82]. When J is odd ($d = J+1$ is even), it is sometimes possible to increase d by unity by appending a parity check which is the sum of all the information bits in the array. This effectively gives J+1 orthogonal parity checks on each information bit, thus permitting $(J+1)/2$ errors to be corrected.

Another answer to the first problem is to make the parity checks more efficient by folding them round the information bit array, without violating the orthogonality conditions. This is equivalent to combining the diagonal checks so that they no longer only cover information bits lying in a straight line. Some quite efficient self-orthogonal array codes can be achieved in this way, and they will be discussed in section 3.1 below.

1.4 Array Codes from Multi-Dimensional Information Bit Arrays

An answer to the second problem posed above is to create a more efficient array code by taking single parity checks on information bit arrays arranged in three or more dimensions, see Figure 5. In the three-dimensional case, the code parameters are $n = n_1 n_2 n_3$, $k = (n_1-1)(n_2-1)(n_3-1) = k_1 k_2 k_3$, and $d = 2x2x2 = 8$. If $n_1 = 5$, $n_2 = 4$ and $n_3 = 3$, then a (60, 24, 8) array code with rate is formed, which is more efficient than the codes with the same or similar values of k mentioned in the previous section. Note, however, that there are only 3 orthogonal sets of parity checks (horizontal, vertical and normal), so that simple majority logic decoding cannot be used. Decoding methods appropriate for product codes, of which this array code is an example, can be used. Also note the important role that the "checks on checks" play in determining the distance of these multi-dimensional array codes. Codes of distance 7, 6, 5 and 4 result if the "check on checks on checks" bit, and 1, 2 or 3 check bit edges are successively removed from the array code cube.

1.5 Burst Error Control with Array Codes

In all the random error control array codes considered so far, the order in which the information and check bits are read out of the encoding array for transfer to the channel is not important, except perhaps if a systematic code is required. The RAC code, for example, can be read out by rows or columns (non-systematic), or by information bit rows or columns followed by the parity row and column (systematic). In the case of burst error control, the read-out order is crucial in determining the properties of the array code.

Any single burst error pattern of length b in a block can be optimally detected by means of b interleaved SPC codes. Therefore a row-parity-check code with parameters $(n_1 n_2, n_1(n_2-1), 2)$, see Figure 6, is an optimum array code capable of detecting single error bursts, or correcting single erasure bursts of length $b \leqslant n_1$, provided the bits of the code are read out by consecutive columns, in the order shown in Figure 6.

By extension of the above concepts, burst error correction requires two orthogonal sets of parity checks, both orthogonal to the read-out order (i.e., the burst direction). Figure 7 shows such an arrangement: an array code with vertical and diagonal parity checks, and horizontal read-out. This code will correct all single bursts of length $b \leqslant i$, but it is awkward that there is a row of $i+j-1$ diagonal checks, compared with the other i-bit rows. By joining (folding sound) the diagonals to form helical parity checks, the code of Figure 8 is obtained. This is an array code with parameters $((i+2)j, ij)$, $i \leqslant j$, capable of correcting an error burst which is confined to one of the $i + 2$ rows of the code (or, alternatively, of correcting two erasure bursts, in different rows). Determining the length of burst which such an array code can correct is, in the general case, an unsolved problem. If i, and the ratio of j to i, is large enough then $b_{row} \leqslant j-1$ (since two bursts of length j can have the same syndrome). When $j = i$, then $2 \leqslant b_{row} \leqslant j-1$, for $j > 2$. Decoding of the row-burst-correcting (RBC) array code consists of aligning, cyclically, the vertical and diagonal syndrome patterns (parity check failures) and then correcting those information bits for which both checks have failed.

When the information bit array is an ixi square then the RBC array code can be re-arranged as shown in Figure 9 for $i = 5$ $(b \leqslant 4)$; because the row order of bit read-out extends into the right-hand check column, in general the value of $b_{row} \leqslant j-1$ stays the same as for the unmodified code. If the read-out of the code in Figure 9 is re-arranged into the diagonal (helical) pattern of Figure 10, then the diagonal checks in 12 become vertical checks in 13, and the vertical checks become horizontal checks. The latter can be shifted out to the right-hand column, and the bottom right-hand bit becomes a check on checks. So the result is a square RAC code, with diagonal read-out, capable of correcting single bursts of errors, or double bursts of erasures, which are confined to the helical pattern diagonals. Using the previous notation, this diagonal-burst-correcting (DBC) array code (phased-burst- correcting code) has parameters $n = n_0^2$, $k = (n_0-1)^2$ and $b_{diag} \leqslant n_0 - 1$. Columns consisting of information bits and a vertical parity check can be added to a DBC code to give an $(n_1 n_2, (n_1-1)(n_2-1), b \leqslant n_1-1)$ code. It is interesting to note that the diagonal read-out of these codes is also the most effective way of logically aligning cells in semiconductor memories [69].

It is a possible disadvantage of both these RBC and DBC codes that they cannot correct phased bursts up to the length of the row or main diagonal, respectively. If the diagonal checks, in the RBC, and the vertical checks, in the DBC, are enhanced by checking additional information bits in the array, then the disadvantage can be overcome. Enhancement of the RBC generates the so-called b-adjacent error correcting code [10]. Such b-adjacent codes (enhanced RBC array codes) exist with information bit arrays of size ixj, $i \leq j-1$, and parameters $((i+2)j, ij, b_{row} = j)$. In a similar way, the vertical parity checks (excluding the vertical check on checks, which is left unaltered) of a DBC array code can be enhanced to generate another type of b-adjacent code. An example is the Patel-Hong code [65,80]. In general, these enhanced DBC array codes have parameters $(n_1(n_1+1), (n_1-1)n_1, b_{diag} = n_1)$, and can also correct any pair of row erasures.

Figure 7 shows a burst-correcting (BC) array code. Its disadvantage is the long row of i+j-1 diagonal check bits. Even though these could be absorbed into the array, it turns out that this is an inefficient way of doing burst correction with an array code. More efficient array codes will be described in section 2.2 below. These BC codes are DBC array codes, as described above, with n_2 sufficiently larger than n_1.

1.6 Shortening of Array Codes

Shortening of random error control array codes is straightforward, because they can be generated in systematic form. The burst error control array codes are in general non-systematic, so shortening is more complicated [7]. The complication arises when a parity check is reached in the codeword read-out order. The solution is to use the idea, previously introduced, of shifting a parity check, by exchanging it with an information bit from the same SPC set. The information bit can then be forced to zero; i.e., shortened from the codeword. If this is not possible, such as when a check on checks bit is involved, then an appropriate information bit elsewhere in the array is forced to a value such that the check on checks is forced to zero; this effectively changes the information bit into a check bit. Check bits which become zero because they have no information bits left to check can be punctured provided that they are consecutive with the bits already shortened; if not, they must remain in the codeword to preserve the burst control properties of the code. Zhang and Wolf [86] have also investigated shortening of burst-correcting array codes.

1.7 Moving Parity Array Codes

All the code considered so far in this section have been block codes. Moving-parity or sliding-parity codes [75,84] are convolutional array codes. An example of such a code is shown in Figure 11 [26]. Each bit in the parity row checks information bits in the "leaning-V" pattern indicated. All the rows are semi-infinite in length, and can be thought of as tracks (e.g., on magnetic tape), with the bits travelling together from right to left while the "checking area" stays fixed (e.g., as a read-head would). The code has rate 4/5 and the constraint length is 5 segments. By analysing the trellis diagram of this convolutional code, it can be shown that it has $d_{free} = 3$, so any single error in a constraint length of the array ($(i+1)^2$ bits = i + 1 segments) can be corrected, and all double errors detected. A very large number of other error patterns are also detected; the exception being, of course, coded sequences in the code. Examples of undetectable error patterns (code sequences) are shown in Figure 11. Note that some of these sequences are subsets of the checking area turned through 180°. The serial free burst-5 distance of this code is $d_{b5free} = 2$, so all bursts running in the vertical dimension, across the tracks (column read-out) of length $\leqslant 5$, are detected; or erasure bursts of length $\leqslant 5$ can be corrected. In the parallel sense, the code will detect any single row (track) burst of errors (correct any single row erasure burst) of any length.

If the arms of the "V" are separated, as in Figure 12, then single row (track) bursts (or double row erasures) of length $\leqslant \ell$ ($\leqslant 4$ in the example) can be corrected. Provided the burst length does not exceed ℓ, then from the number of bits between the repetitions of the burst error pattern as seen by the arms of the "V", the row that the burst is in can be determined, and hence it can be corrected. The error-free guard space required between bursts is L (= 8 in the example) = ℓ+i. A variant of this burst-correcting moving parity code uses a symmetrical V checking area, with both arms checking diagonally (at +45°) across the tracks [25,30].

2. BLOCK ARRAY CODES

This section looks in more detail at three particular types of block array code: self-orthogonal and related random-error-correcting array codes; burst-error-correcting array codes; and array codes for correcting clusters (two-dimensional bursts) of errors.

2.1 Self-Orthogonal and Related Random-Error-Correcting Array Codes

A pxp information bit array, p a prime, has rows and columns labelled 0, 1,..., r,...,p-1, and 0, 1,...c...,p-1, respectively. For each information bit, d-1 numbers $D_i(r,c) = (r+ic) \bmod p$, $0 \leqslant i \leqslant d-2$, $i < p$ (i.e., $d \leqslant p+1$), are calculated. Smith [81] proved that no two information bits have more than one common value of $D_i(r,c)$. Letting each value which each $D_i(r,c)$ takes define a parity check covering the corresponding information bits, the result is an (n, p^2, d) self-orthogonal random-error-correcting array code, where $n = p^2+p(d-1)$. The number of check bits is p(d-1) because the values of $D_i(r,c)$ determine d-1 sets of p parity checks. Some of these sets will be row (or column) and diagonal checks; the others are "folded" checks. If a parity check over all the information bits is appended to the codeword, then distance 5 is achieved and double error correction is possible, as indicated in section 1.3 above. This enhancement has been generalised by Horiguchi and Morita [48]. These binary self-orthogonal array codes can achieve high rates, with parameters close to, and sometimes better than, the best self-orthogonal quasi-cyclic codes [83]. They are very easily decoded using a 1-step majority logic algorithm [55]. The construction also works when p is a power of a prime, though the decoders are more complex.

The above construction can be generalised [73] by using a three-dimensional information bit array, with parameters pxpxh. In the simplest case, h = 2 (h = 1 is the Smith construction). Each information bit is now associated with d-1 numbers given by $D_i(r,c,\ell) = (r+ic+i^2\ell)$ where r, c and ℓ label the rows, columns and layers of the array, respectively, $\ell = 0$ or 1, $0 \leqslant i \leqslant d-2$, and $i < p$ ($d \leqslant p+1$). Associating parity checks with $D_i(r,c,\ell)$ values, as before, generates codes with rates $\geqslant 2/3$ (e.g., for p = 3 and d = 4, the parameters are (27, 18, 4)). Array codes with this construction are not self-orthogonal, and hence are not one-step majority logic decodable, because some pairs of information bits have more than one common value of $D_i(r,c,\ell)$. Decoding is therefore rather more complex, but the codes are more efficient, than those of the Smith construction. Codes with larger values of $h \leqslant p$ are also possible, but apart from not being self-orthogonal, it is not clear whether the design distance is achieved in all cases.

Random-error-control array code constructions can be based on the incidence matrices of block designs [42] and on the so-called complex-rotary technique [33,34,18]. Jin Fan [35] has also shown that many of the complex-rotary constructions are equivalent to block design constructions. For example, a class of such codes with parameters $(2p^2+p, p^2, p+2)$, p a prime, can correct (p+1)/2 errors with one-step majority logic decoding. This class is based on the balanced incomplete block design (BIBD) with parameters $(v,b,r,k,\lambda) = (p^2, p(p+1), p+1, p, 1)$. Similar equivalences exist between array code classes generated from differential balanced block designs (DBBD) or symmetrical BIBDs (SBIBD), and extended complex-rotary code classes. A class with parameters $(2(p^2+p+1), p^2+p+1, p+2)$, which can be derived from either the $(v,k,\lambda) = (p^2+p+1, p+1, 1)$ SBIBD or the extended complex-rotary construction, are easily decoded. The constructions also work when p is a power of a prime; in this case the class of one-step majority logic decodable codes derived from orthogonal Latin squares [49] is included.

A wing array code [45,56] consists of a right-angled triangle of information bits with a row of parity checks along its base, the whole forming a wing as in Figure 13. The checks cover the information bits along lines at $\pm 45^o$ to the parity row. A single-error-correcting code with parameters (r(r+1)/2, r(r-1)/2, 3) is the result, where r is the number of parity checks. Multiple-wing SEC codes can be obtained by attaching further information wings to the same set of parity checks. Each information wing must have a parity bit checking all the information bits in the wing also attached to it, to enable the wing with the error in it to be determined. This construction yields an ([rs(r-1)/2]+r+s, rs(r-1)/2, 3) code, where s is the number of wings. Multiple-wing codes may be further combined to yield checkerboard codes with parameters (ab, a+b-1, min {a,b}).

2.2 Burst-Correcting Array Codes

This section describes several classes of single-burst-correcting RAC array codes with diagonal read-out (DBC array codes). Multiple-burst-correcting codes developed from these codes are also described. Almost all the codes described below correct end-round (cyclic) bursts.

Consider an $n_1 x n_2$ ($n_2 \geqslant n_1$) binary array code, in which each row and column has even weight (an RAC code). Diagonal (helical) read-out commences from the top left-hand corner, runs down the main diagonal, jumps to the diagonal s ($1 < s \leqslant n_2-1$) places to the right of the main diagonal, and so on. An example for $s = 3$ is given in Figure 14. n_2 and s must be relatively prime so that all bits in the codeword are covered once. Codes of this construction are capable of correcting single bursts of length $b \leqslant n_1$. The exact value of b is a function of n_1, n_2 and s. These codes were first studied by Gilbert [43] and others [2,11,58] who used $s = n_1$, and therefore n_1 and n_2 are required to be relatively prime. In this case the codes are cyclic, and for some values of n_1 and n_2 are equivalent to Fire [40] codes. The maximum length of burst that Gilbert codes can correct was only known in some cases, until Zhang and Wolf [86] were able to determine b exactly. They proved that, for any appropriate value of s, $b \leqslant n_1-s$ if $1 \leqslant s \leqslant n_1-1$, and $b \leqslant n_1$ if $n_1 \leqslant s \leqslant n_2-1$. They also were able to determine the value of s which gives the maximum value of b for any particular combination of values of n_1 and n_2, and they provide a large table of these optimum burst correcting array code parameters.

The case $s = 1$ was investigated by Farrell and Hopkins [38, 39]. Independently, Campello de Souza [61] and Blaum, et. al., [6], proved that $b = n_1-1$, provided $n_2 \geqslant 2n_1-3$. There are many fast and easy ways of decoding these codes. The conceptually simplest method is to locate the unique run of $\geqslant n_1-2$ zeros in the vertical syndrome (viewed cyclically). This allows the corresponding non-zero entries of the horizontal and vertical syndrome to identify the burst error pattern [6]. A simple error-trapping decoder can also be used [38]. For these codes, $n-k = n_1+n_2-1 = 3n_1-4$ when $n_2 = 2n_1-3$. Since $2b = 2(n_1-1) = 2n_1-2$, the codes do not meet the Reiger [71] burst-correction burst-correction bound. Letting $z = 2b/(n-k) = (2n_1-2)/(3n_1-4)$ then $z \to 2/3$ as $n_1 \to \infty$. This result is also true for Fire codes (see [52]). The case $s = n_2-1$ (abbreviated $s = -1$) is interesting because it is one of at least two values of s which permits $b = n_1$; for $s = -1$, $b = n_1$ iff $n_2 \geqslant 2n_1+1$ [7]. Once again there is a fast and simple decoding algorithm. The number of parity bits is $n-k = 3n_1$ when $n_2 = 2n_1+1$; so $z \to 2/3$ as $n_1 \to \infty$. The case $s = -2$ ($= n_2-2$) also permits $b = n_1$, provided $n_1 = 4u+v+2$ and $n_2 = 6u+2v+5$, $u \geqslant 1$, $v \geqslant 0$ but $v \neq 1$ [5]. A slightly more complicated, but still relatively fast, decoding algorithm exists for these codes. Here $n-k = 10u+3v+6$ and $b = n_1 = 4u+v+2$, so $z \to 4/5$ as $u \to \infty$.

Now consider an $n_1 x n_2$ ($n_2 \geqslant n_1$) binary array code which has rows consisting of codewords from an (n_2, k_2, $d=2t$) code, and columns with even parity (e.g., the column code has parameters (n_1, n_1-1, 2)). Diagonal read-out with parameter s is used as before. Daniel [22,23] devised array codes of this type with parameters $t = 2$ (extended Hamming SEC codes with $d = 4$) and $s = 1$. These codes were constructed heuristically, by manipulating the parity-check matrix of the extended Hamming row code to ensure double-burst correction. In most cases $b = n_1-1$ was achieved but in some cases only $b = n_1-2$ was possible, because n_2 was too small. A moderately complex decoding algorithm was devised, which depended on the particular structure of the row code. These codes were all capable of operation as single-burst-detecting codes with $b = 2n_1-1$.

These codes were effectively generalised by Blaum, et. al., [7], who proved that with parameter $s = 1$, the array code defined in the previous paragraph can correct up to t bursts of length $\leqslant n_1-1$ iff $n_2 \geqslant 2t(n_1-2)+1$. When $s = -1$, the code can correct up to t bursts of length $\leqslant n_1$ iff $n_2 \geqslant 2tn_1+1$. In both cases, array row codes of distance 2t can be used; and a moderately complex general decoding algorithm exists for the array code. It is interesting to note that in some cases the values of n_2 reported by Daniel [22,23] are less than the minimum value of $2t(n_1-2)+1$, because a specific row code is used.

Another general technique for constructing multiple-burst-correcting (MBC) array codes is to take any t-burst correcting code as row code, with the SPC code as column code. This array code will correct up to 2t bursts [23]. This property comes from the product properties of the array code: if the row code has burst distance [12] $d_{bb} = 2t+1$, then the array (product) code has $d_{bb} = d_1.d_{bb2} = 2(2t+1) = 4t+2$; i.e., the array code can correct $\lfloor(4t+1)/2\rfloor = 2t$ bursts of length b. The row code may itself be an array code, of course, in which case a three-dimensional code results. A fairly simple decoding algorithm has been devised for these three-dimensional codes [23]; alternatively, a modified form of the Bloch-Zyablov algorithm for concatenated codes can be used [9,32].

Bridwell and Wolf [12] have studied the multiple-burst-correction properties of two-dimensional product codes, constructed from row and column component codes (not necessarily cyclic) with parameters (n_1, k_1, d_{bb1}) and (n_2, k_2, d_{bb2}). Gilbert $(s = n_1)$ diagonal red-out is used, which they attribute to Abramson (unpublished); as well as row read-out. The minimum burst-b distance for such product codes is determined, and a class of multiple-burst correcting cyclic product codes presented. Use of the burst-b distance concept makes their results very general and widely applicable.

2.3 Cluster-Correcting Array Codes

Clusters, patches or two-dimensional bursts of errors occur in many digital transmission, processing or storage systems, wherever data is formatted in two dimensions (e.g., magnetic tape, etc.). Array codes are very suitable for correcting cluster errors in such two-dimensional data structures.

The simplest form of single cluster correction code is the interleaved RAC code [37]. An ixi cluster of e errors, $1 \leqslant e \leqslant i^2$, in an array code of parameters $(k_1+i)(k_2+i)$, with a $k_1 x k_2$ information bit array, can be corrected by interleaving both row and column checks to depth i. Note that the ixi (2x2) checks on checks are not essential, and can be punctured if convenient. Decoding is fast and simple as for the RAC code. Clusters can also be rectangular rather than square, in which case the interleaving depth will be different for rows and columns.

Single-burst-correcting array codes, as described in section 2.2 above, interleaved by rows to depth i, can be used to correct a single cluster of ixi errors, $1 \leqslant e \leqslant i^2$ [37]. Column interleaving is also possible. Almost any desired shape of array can be constructed. Suitable RBC or DBC array codes may be useful for constructing relatively broad and thin arrays. These codes are more efficient than interleaved RAC codes.

3. CONVOLUTIONAL ARRAY CODES

Several types of convolutional array code are mentioned in this section, including orchard codes, MDS codes and cross-interleaved codes.

3.1 Orchard Codes

This is a generic name that can be given to all those convolutional array codes which are of the moving or sliding parity type introduced in section 1.7 above, including both random and burst error correcting codes. Almost every known class of convolutional codes can be interpreted as an orchard code in some sense, but perhaps the earliest important class of orchard codes is the self-orthogonal recurrent codes, based on difference triangles, constructed by Robinson and Bernstein [72], with rates 1/n or n-1/n. They all can be majority logic (threshold) decoded [55] in a fast and simple way. Scott and Goetschel [75] also constructed orchard codes based on difference triangles; a rate 2/3, double-error-correcting, example is given in Figure 15. Note that previously computed parity checks are included in the current checking pattern; Otter [60] regards this as the distinguishing feature of orchard codes proper, a narrower definition than the one given above. These codes are not necessarily self-orthogonal, so decoding is more complex. Shiozaki [77] has proposed orchard codes constructed such that the parity pattern of each row is given by the polynomial $1+x^{n+1-i} + x^{n+i}.g(x)$, where n is the number of rows, i is the row number $(1 \leqslant i \leqslant n)$ and g(x) is a primitive polynomial of degree r satisfying $n \leqslant 2^r-1$. A similar type of orchard code, with rate 5/6 and also double-error-correcting, is shown in Figure 16 [77]. Otter [60], Otter and de Vries [61], Lee, et. al., [51] and Shiozaki [78] have devised a number of orchard codes and characterised their properties.

Many moving parity (orchard) schemes have been devised for the correction of bursts on magnetic tapes [13,62,64,68,69,84]. These are often called cross-parity codes, where the "cross-parity" refers to diagonal checks with positive (and also sometimes negative) slopes. Figure 17 shows a simple cross-parity code [28,84]. It is a moving parity version of the RBC block array code described in section 1.5 (Figure 7), except that the diagonal checks also check the vertical check bits (which makes it a narrow-sense orchard code). Figure 18 is a diagram of the binary Patel [64] rate 7/9 code; this adaptive cross parity code has minimum distance $d_{min} = 4$ (minimum track weight over the constraint length), but is capable of correcting most patterns of up to four row (track) erasure bursts, two row erasure bursts and one row error burst, or two error bursts. Note that the rows of the format in Figure 18 are interleaved in practice, and that this code is a combination of two of the simple "V" burst correcting moving parity codes described in section 1.7 above. Alternative vertical and serial byte-oriented formats for the diagonal checks are also possible (Patel, [64]). Decoding is of only moderate complexity.

3.2 Maximum Distance Separable Orchard Codes

It is of interest to find optimum orchard (moving parity) codes, with respect to rate and minimum distance. Maximum distance separable (MDS) orchard codes [4,62,66,68] have $d_{min} = n-k+1$, which is the maximum possible distance for a code with given n (number of rows) and k (number of information rows). The most useful class of MDS block codes is the Reed-Solomon codes: Blaum's rate 12/18 = 2/3 MDS orchard code has $n-k = 6$, $d_{min} = n-k+1 = 7$, so it can correct up to 6 track (row) erasure bursts, and other combinations of track erasure and error bursts within the distance (e.g., two error bursts and two erasure bursts; or three error bursts). Erasure-only decoding is particularly simple to implement. Fuja [41] has also studied MDS moving parity codes, which he calls cross-parity-check (CPC) codes.

3.3 Cross-Interleaved Codes

One way of encoding orchard codes (convolutional array codes) with row and diagonal checks [68] is to use two stages of vertical parity checking with a stage of convolutional interleaving in between. Figure 19 shows an example, for four information rows and two parity rows. The combination is called cross-interleaving [67,84] and is a convolutional orchard code. Note that if the encoded stream is truncated, and the diagonal (vertical) checks are folded round as in Figure 8, then the familiar row burst correcting code emerges. This truncating process (tail-biting), when applied to cross-interleaved codes, is sometimes called block-completed convolutional interleaving [84]. Cross-interleaved codes may be decoded in an integrated process, or the two component codes can be decoded separately, as in a normal concatenated coding scheme.

4. BYTE-ORIENTED ARRAY CODES

All the binary parity checking array codes described in the previous sections generalise to array codes with symbols over GF(q), where q is a power of a prime. If $q = 2^m$, then an m-bit byte can be thought of as an extra dimension, lying "behind" each bit in the arrays constructed for binary error control. Thus a two-dimensional binary RAC code becomes a three-dimensional q-ary RAC code, etc.. Bytes also can be accommodated in any suitable format within the dimensions of the binary array code. When an array code is composed of one or more codes which are not SPC codes, then some of the array code parameters may need to be modified. Single symbol correction in $GF(2^m)$ is possible with two check symbols, using a Reed-Solomon (RS) code [53,70] whereas a binary Hamming SEC code requires at least three check bits [44]. All the above assumes that symbol errors occur, on the channels that these byte-oriented array codes are intended for. It is also possible to modify binary array codes to provide efficient protection against binary channel errors (e.g., in phased or un-phased bursts) for data organised into byte format (often already including an SPC within the byte (e.g., an (8, 7, 2) code)). Some particular byte-oriented codes, block and convolutional, are mentioned below.

4.1 Block Byte-Oriented Array Codes

Seguin, et. al., [76] have devised an array code for 8-bit bytes consisting of (8, 7, 2) SPC codes. It can correct single errors, correct double errors occurring in two different bytes, and detect all other double errors. Courteau and Goulet [21] have extended these codes and devised a decoding algorithm based on the geometric properties of the code. Other extensions and generalisations have been devised by Mortimer, et. al., [57], Harbour [46] and Chen, [17].

A code which combines bit and byte error control is described by Patel [63]. It uses the (15, 13, 3) BCH code [52] with symbols in $GF(2^4)$, but replaces these symbols by the 16-symbol sub-field of $GF(2^8)$ for error protection purposes. This makes it appropriate for 8-bit byte information; the code can correct all single bit and byte errors, and detect all double bit errors (in different symbols). The code is interleaved and provided with overall parity and a sync. word in each row of the interleaver; this is of additional interest in the context of array codes.

The 9x8 Patel-Hong [65] code has already been introduced in section 1.5 above. This code was originally devised as a byte-oriented code, capable of correcting one symbol error (or two symbol erasures) in the rows (tracks) of the codeword. Rows and columns of the code are symbols in $GF(2^8)$, and the irreducible polynomial $g(x) = 1+x^3+x^4+x^5+x^8$ is used to generate the elements of $GF(2^8)$. The bottom row is a symbol SPC in $GF(2^8)$, and the left-hand column (excluding the parity check) is the sum of the other columns (excluding the checks) multiplied by x^i, where i is the column position $1 \leq i \leq 7$ ($i = 0$ is the redundant left-hand column) in $GF(2^8)$. Blaum and McEliece [8] generalised this code to make it able to correct three erasure rows (or an error row and an erasure row).

4.2 Convolutional Byte-Oriented Array Codes

Convolutional moving parity array codes with vertical and diagonal checks, as mentioned in section 3.2 above, can be generalised by using vertical and diagonal byte-oriented symbol correcting codes. An example is the scheme of Cheung [19], which has 17 rows (tracks) of byte-oriented data, 4 of which are parity rows. Two parity rows at the top of the array provide diagonal coding; two at the bottom provide vertical coding. The component code is the (15, 13, 1) BCH code over $GF(2^8)$. This is the same code as used by Patel [63], but by proper choice of the irreducible polynomial defining the elements of $GF(2^8)$, Cheung's scheme is able not only to correct single bit or symbol (byte) errors, and detect two bit errors (in different symbols), but it can also detect a single error in one symbol and a burst of 2 bit errors in another, or two bursts of 2 bit errors (in different symbols). This scheme is called a hybrid code by Cheung because decoding combines the properties of block and convolutional codes.

The above scheme [19] is an example of cross-interleaving. A very powerful error-correcting scheme consists of cross-interleaving two RS codes, as used in compact discs [27,29,50,67,69,84]. Several decoding strategies are possible here, ranging from independent decoding of each component code, as in a concatenated scheme, to interactive decoding involving the column code marking erasures and making tentative decoding estimates for the diagonal decoder to process and feed back on, in an iterative multi-pass decoding operation. Ko and Tjhung [50] compare the performance of some of these strategies. Shiozaki [78] has studied byte-oriented orchard codes.

5. CONCLUSION

Array codes offer a very wide range of flexible error control characteristics for many applications. They are relatively simple to design, encode and decode. Some array codes (e.g., the maximum distance separable orchard codes) are optimum in the sense of requiring the minimum number of parity checks for their error-control power and the number of information bits protected. Other array codes are not optimum in this sense, but offer particularly high-speed, low complexity decoding in return (e.g., the self orthogonal and burst-correcting array codes). Within the class of array codes, the various types offer trade-offs between higher efficiency (e.g., rate) and decoding complexity. On the one hand, all the desirable error-control functions (error detection, error and/or erasure correction, of single and multiple random, burst or cluster patterns) can be provided by quite simple two-dimensional row-and-column parity array codes with diagonal (helical) read-out. Decoding complexities range from low to moderate, but a penalty in some cases is the relatively low ratio of error pattern size (length) to array code size (block length). On the other hand, there are lower rate and shorter block length array codes, such as certain convolutional and byte-oriented types (e.g., cross-interleaved RS codes) which require moderate to high complexity decoding. All array codes are conceptually simple to understand and can be combined in a multiplicity of useful ways.

REFERENCES

1. L. R. Bahl and R. T. Chien: On Gilbert Burst-Error-Correcting Codes; IEEE Trans., Vol. IT-15, N. 3, pp. 431-433, May 1969.
2. L. R. Bahl and R. T. Chien: Single- and Multiple-Burst-Correcting Properties of a Class of Cyclic Product Codes; IEEE Trans., Vol. IT-17, No. 5, pp. 594-600, Sept. 1971.
3. E. R. Berlekamp: Algebraic Coding Theory; McGraw-Hill, 1968.
4. M. Blaum: A Family of Error-Correcting Codes for Magnetic Tapes; IBM Research Report, 1985.
5. M. Blaum: A Family of Efficient Burst-Correcting Array Codes; IBM Res. Report, March 1989.
6. M. Blaum, P. G. Farrell and H. C. A. van Tilborg: A Class of Burst-Error-Correcting Array Codes; IEEE Trans., Vol. IT-32, No. 6, pp. 836-839, Nov. 1986.
7. M. Blaum, P. G. Farrell and H. C. A. van Tilborg: Multiple-Burst-Correcting Array Codes; IEEE Trans., Vol. IT-34, No. 5, pp. 1061-1066, Sept. 1988.
8. M. Blaum and R. J. McEliece: Coding Protection for Magnetic Tapes: A Generalisation of the Patel-Hong Code; IEEE Trans., Vol. IT- , No. 5, pp. 690-693, Sept. 1985.
9. E. L. Blokh and V. V. Zyablov: Generalised Concatenated Codes; Svyaz, Moscow, 1976.
10. D. C. Bossen: b-Adjacent Error Correction; IBM Jour. Res. and Dev., Vol. 14, pp. 402-408, 1970.
11. R. T. Bow: Codes for High Speed Arithmetic and Burst Correction; Ph.D. Thesis, University of Illinois at Urbana-Champagne, USA, Jan. 1973.
12. J. D. Bridwell and J. K. Wolf: Burst Distance and Multiple-Burst Correction; BSTJ, Vol. 49, pp. 889-909, May-June, 1970.
13. D. T. Brown and F. F. Sellers: Error Correction for IBM 800-bit-inch Magnetic Tape; IBM Jour. Res. and Dev., Vol. 14, pp. 384-389, 1970.
14. H. O. Burton and E. J. Weldon: Cyclic Product Codes; IEEE Trans., Vol. IT-11, pp. 433-439, July 1965.
15. P. Calingaert: Two-Dimensional Parity Checking; Jour. Assoc. Comp. Machinery, Vol. 8, No. 2, pp. 186-200, 1961.
16. R. Campello de Souza: Single-Burst Correcting Array Codes; Comms. Res. Group Report, University of Manchester, 1984.
17. C. L. Chen: Byte-Oriented Error-Correcting Codes for Semiconductor Memory Systems; IEEE Trans., Vol. C-35, No. 7, pp. 646-648, July 1986.
18. Chen Ji: Extended Complex-Rotary Codes and their Encoding/Decoding Methods; Dissertation, South-Western Jiaotong University, 1986.
19. Kar-Ming Cheung: Error-Correction Coding in Data Storage Systems; Ph.D. Thesis, Calif. Inst. of Tech., USA, 1987.
20. A. B. Cooper and W. C. Gore: Iterated Codes with Improved Performance; IEEE Trans., Vol. IT-24, No. 1, pp. 116-118, Jan. 1978.

21. B. Courteau and J. Goulet: An Extension of the Codes Introduced by Seguin, Allard and Bhargava; Discrete Maths., Vol. 56, pp. 133-139, 1985.

22. J. S. Daniel: Double-Burst-Correcting Array Codes: Generation and Decoding; IEEE Int. Symp. on Info. Theory, Brighton, UK, June 1985.

23. J. S. Daniel: Synthesis and Decoding of Array Error Control Codes; Ph.D. Thesis, University of Manchester, UK, August 1985.

24. J. S. Daniel and P. G. Farrell: Burst-Error-Correcting Array Codes: Further Developments; IERE Conf., Dig. Proc. of Signals in Comms., Loughborough, April 22-26, 1985.

25. N. Darwood: A Moving Parity Check Method; Electronic Eng., April 1979.

26. N. Darwood: Improved Parity Checker; Wireless World, pp. 81-82, Jan. 1981.

27. L. B. de Vries and K. Odaka: CIRC - The Error Correcting Code for the Compact Disc Digital Audio System; Digital Audio Collected Papers, AES, 1983.

28. T. T. Doi: Channel Coding for Digital Audio Recordings; Jour. Audio Eng. Soc. Vol. 31, No. 4, pp. 224-238, April 1983.

29. T. T. Doi, K. Odaka, G. Fukuda and S. Furukawa: Cross-Interleave Code for Error Correction of Digital Audio Systems: Jour. Audio Eng. Soc., Vol. 27, pp. 1028- , 1979.

30. T. Donnelly: Real Time Microprocessor Techniques for a Digital Multitrack Tape Recorder; Ph.D. Thesis, Plymouth Polytechnic (CNAA), Jan. 1989.

31. P. Elias: Error-Free Coding; IEEE Trans., Vol. IT-4, pp. 29-37, 1954.

32. T. Ericsson: A Simple Analysis of the Blokh-Zyablov Decoding Algorithm, Internal Report, Linkoping University, Sweden, Nov. 1986.

33. Jin Fan: An Investigation on New Complex-Rotary Codes; IEEE Int. Symp. on Info. Theory, Brighton, UK, June 1985.

34. Jin Fan: Analysis of the Characteristics of a New Complex Rotary Code; Jour. of Chinese Inst. of Comms., Vol. 7, No. 2, 1986.

35. Jin Fan: A New Combinatorial Coding Method using Block Design; Proc. 1988 Beijing Int. Workshop on Info. Theory, July 4-7, 1988.

36. P. G. Farrell: Array Codes; Algebraic Coding Theory and Applications, Ed. G. Longo, Springer-Verlag, pp. 231-242, 1979.

37. P. G. Farrell: Array Codes for Correcting Cluster-Error Patterns; IEE Conf. on "Electronic Signal Processing", York, July 26-28, 1982.

38. P. G. Farrell and S. J. Hopkins: Burst-Error-Correcting Array Codes; Radio and Electronic Engr., Vol. 52, No. 4, pp. 182-192, April 1982.

39. P. G. Farrell and S. J. Hopkins: Decoding Algorithms for a Class of Burst-Error-Correcting Array Codes; IEEE Int. Symp. on Info. Theory, Les Arcs, France, June 21-25, 1982.

40. P. Fire: A Class of Multiple-Error Correcting Binary Codes for Non-Independent Errors; Sylvania Rep. RSL-E-2, April 24, 1959.
41. T. Fuja: On the Structure and Decoding of Cross-Parity-Check Codes for Magnetic Tape; Ph.D. Thesis, Cornell University, May 1987.
42. B. J. Gassner: Cyclic BIB Designs; IEEE Int. Conv. Record, Pt. 7, p. 259, 1965.
43. E. N. Gilbert: A Problem in Binary Encoding: Proc. Symp. Applied Maths., Vol. 10, Ann. Math. Soc., pp. 291-297, 1960.
44. R. W. Hamming: Error Detecting and Error Correcting Codes; BSTJ, Vol. 29, pp. 147-160, 1950.
45. S. Harari: Fuzzy Correction Capability; Algebraic Coding Theory and Applications, Ed. G. Longo, Springer-Verlag, pp. 497-512, 1979.
46. T. G. Harbour: Burst Error Correcting Codes Based on Modified Array Codes; Personal Communication, June 1987.
47. S. J. Hong and A. M. Patel: A General Class of Maximal Codes for Computer Applications; IEEE Trans., Vol. C-21, p. 1322, 1972.
48. T. Horiguchi and K. Morita: A Parallel Memory with Double Error Correction Capability - A Class of One-Step Majority-Logic Decodable Error Correction Codes; IECE Japan, Tech. Group, EC 75-42, Nov. 1975.
49. M. Y. Hsiao, D. C. Bossen and R. T. Chien: Orthogonal Latin Square Codes; IBM Jour. Res. and Dev., Vol. 14, pp. 390-394, July 1970.
50. C. C. Ko and T. T. Tjhung: Performance of Simple Cross-Interleaved Reed-Solomon Decoding Strategies for Compact Disc Players; Int. Jour. Electronic, Vol. 64, No. 4, pp. 627-635, 1988.
51. J. H. Lee, T. Segawa and A. Shiozaki: A Decoding Method for Shizaki's Orchard Codes and a Proposal of a New Class of Orchard Codes; Proc. 1988, Beijing Int. Workshop on Info. Theory, July 4-7, 1988.
52. S. Lin and D. J. Costello: Error-Control Coding: Fundamentals and Applications; Prentice-Hall, 1983.
53. F. J. MacWilliams and N. J. A. Sloane: The Theory of Error-Correcting Codes; North-Holland, 1977.
54. T. Mano, J. Yamada, J. Inove and S. Nakajima: Circuit Techniques for a VLSI Memory: IEEE Jour. Solid State Circuits, Vol. SC-18, No. 5, pp. 463-469, Oct. 1983.
55. J. L. Massey: Threshold Decoding; MIT Press, 1963.
56. B. Montaron: Codes a Ailerons, Codes a Damiers et Constructions Combinatories; These 3eme Cycle, Univ. de Paris VI, May 1978.
57. B. C. Mortimer, M. J. Moore and M. Sablatash: The Design of a High-Performance Error-Correcting Coding Scheme for the Canadian Broadcast Telidon System Based on Reed-Solomon Codes; IEEE Trans., Vol. COM-35, No. 11, pp. 1113-1123, Nov. 1987.
58. P. G. Neumann: A Note on Gilbert Burst-Correcting Codes, IEEE Trans., Vol. IT-11, No. 4, pp. 377-384, July 1965.

59. F. I. Osman: Error-Correction Techniques for Random-Access Memories; IEEE Jour. Solid State Circuits, Vol. SC-17, No. 5, pp. 877-881, Oct. 1982.

60. E. L. Otter: The Orchard Error-Correcting Codes; Ph.D. Dissertation, University of New Mexico, Dec. 1984.

61. E. L. Otter and R. C. de Vries: A Method of Examining Orchard Codes for Minimum Hamming Distance Five; IEEE Trans., Vol. COM-34, No. 4, pp. 399-404, April 1986.

62. A. M. Patel: Multitrack Error Correction with Cross-Parity Check Coding; IBM Tech. Rep. TR02.813, March 20th, 1978.

63. A. M. Patel: Error Recovery Scheme for the IBM 3850 Mars Storage System; IBM Jour. Res. and Dev., Vol. 24, No. 1, pp. 32-42, Jan. 1980.

64. A. M. Patel: Adaptive Cross-Parity (AXP) Code for the IBM Magnetic Tape Subsystem; IBM Jour. Res. and Dev., Vol. 29, No. 6, pp. 546-562, Nov. 1985.

65. A. M. Patel and S. J. Hong: Optimal Rectangular Code for High Density Magnetic Tape; IBM Jour. Res. and Dev., Vol. 18, pp. 579-588, Nov. 1974.

66. P. Piret and T. Krol: MDS Convolutional Codes; IEEE Trans., Vol. IT-29, No. 2, pp. 224-233, March 1983.

67. K. C. Pohlmann: Principles of Digital Audio; Sams (MacMillan), 2nd Edition, 1989.

68. P. Prusinkiewicz and S. Budkowski: A Double-Track Error-Correction Code for Magnetic Tape; IEEE Trans., Vol. C-19, No. , pp. 642-645, June 1976.

69. T. R. N. Rao and E. Fujiwara: Error-Control Coding for Computer Systems; Prentice-Hall, 1989.

70. I. S. Reed and G. Solomon: Polynomial Codes over Certain Finite Fields; Jour. Soc. Indust. Appl. Maths., Vol. 8, pp. 300-304, 1960.

71. S. Reiger: Codes for the Correction of Clustered Errors, IRE Trans., Vol. IT-6, pp. 16-21, March 1960.

72. J. Robinson and A. Bernstein: A Class of Binary Recurrent Codes with Limited Error Propagation; IEEE Trans., Vol. IT-13, No. 1, pp. 106-113, Jan. 1967.

73. V. C. da Rocha: On an Extension of the Smith Construction for Self-Orthogonal Array Codes; Private Communication, 1988.

74. R. Rowland: Error-Detecting Capabilities of Two-Coordinate Codes; Electronic Eng., Vol. 40, pp. 16-20, Jan. 1968.

75. E. Scott and D. Geotschel: One Check Bit Per Word can Correct Multibit Errors; Electronics, pp. 130-134, May 5th, 1981.

76. G. Seguin, P. E. Allard and V. Bhargava: A Class of High Rate Codes for Byte-Oriented Information Systems; IEEE Trans., Vol. COM-31, No. 3, pp. 334-342, March 1983.

77. A. Shiozaki: Proposal of a New Coding Pattern in Orchard Scheme; Info. and Control, Vol. 51, No. 3, pp. 209-215, Dec. 1981.

78. A. Shiozaki: A New Class of Orchard Codes with Double Bit Error and Single Byte Error Correcting Capability; IECE Japan Trans., Vol. E69, No. 12, pp. 1330-1333, Dec. 1986.

79. D. Slepian: Some Further Theory of Group Codes; BSTJ, Vol. 39, pp. 1219-1252, Sept. 1960.
80. N. J. A. Sloane: A Simple Description of an Error-Correcting Code for High-Density Magnetic Tape; BSTJ, Vol. 55, No. 2, pp. 157-165, Feb. 1976.
81. R. J. G. Smith: Easily-Decoded Efficient Self-Orthogonal Block Codes; Elec. Letters, Vol. 13, No. 7, pp. 173-174, March 31st, 1977.
82. R. J. G. Smith: Easily-Decoded Error-Correcting Codes; Ph.D. Thesis, University of Kent at Canterbury, UK, 1978.
83. R. I. Townsend and E. J. Weldon: Self-Orthogonal Quasi-Cyclic Codes; IEEE Trans., Vol. IT-13, pp. 183-195, 1967.
84. J. Watkinson: The Art of Digital Audio; Focal Press, 1988.
85. J. Yamada, T. Mano, J. Inoue, S. Nakajima and T. Matsuda: A Submicron 1-Mbit Dynamic RAM with a 4-bit-at-a-time Built-In ECC Circuit; IEEE Jour. Solid State Ccts., Vol. SC-19, No. 5, pp. 627-633, Oct. 1984.
86. W. Zhang and J. K. Wolf, A Class of Binary Burst Error-Correcting Quasi-Cyclic Codes; IEEE Trans., Vol. 34, No. 3, pp. 463-479, May 1988.

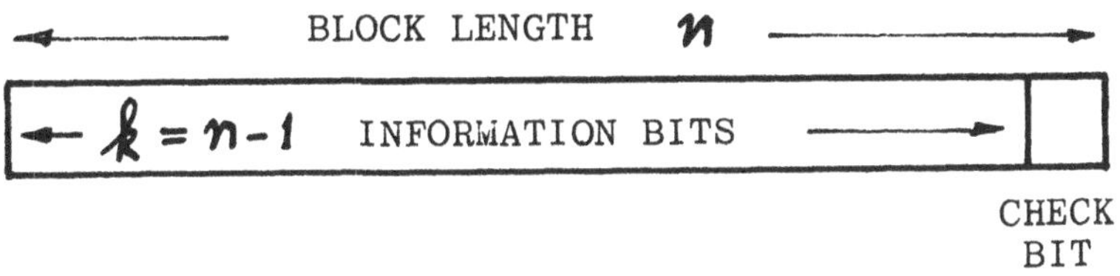

FIGURE 1 : SPC CODE

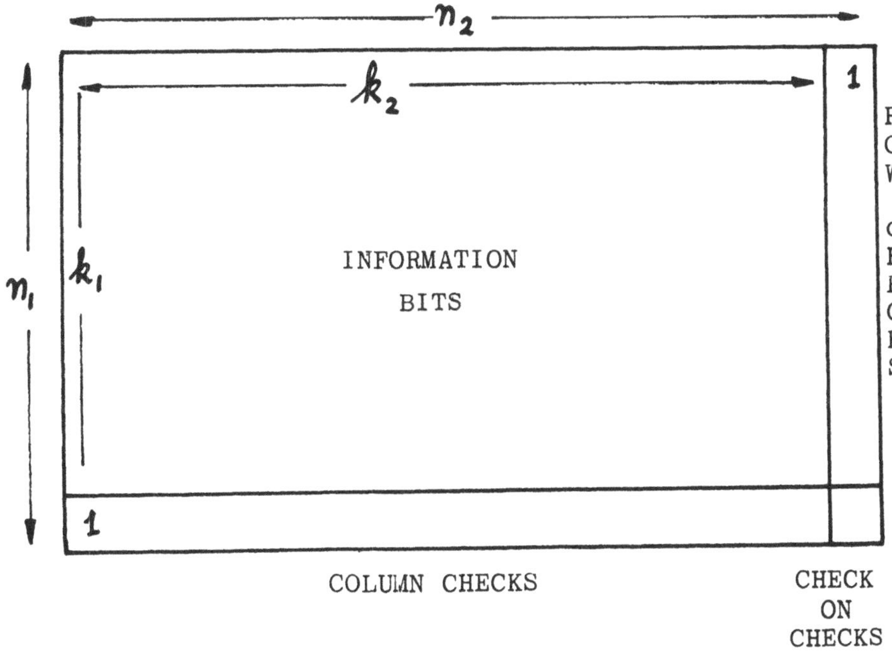

FIGURE 2 : RAC CODE

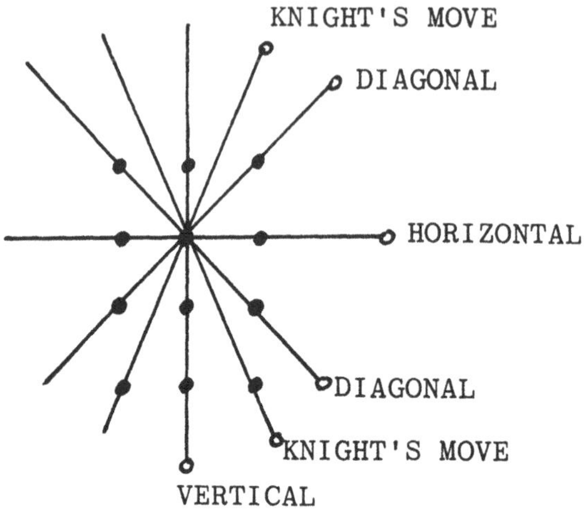

FIGURE 3 : CHECKS IN SIX DIRECTIONS

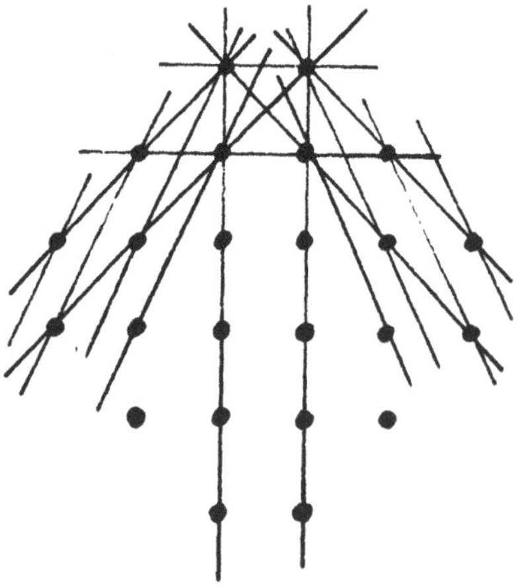

FIGURE 4 : (74,24,7) ARRAY CODE

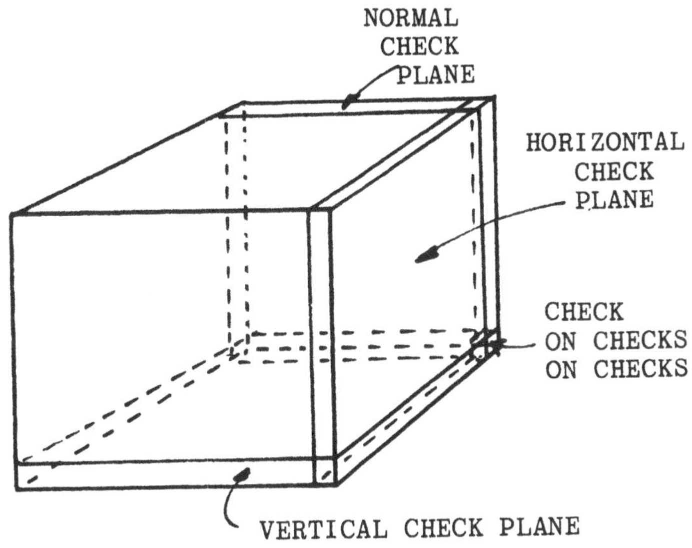

FIGURE 5 : THREE-DIMENSIONAL ARRAY CODE

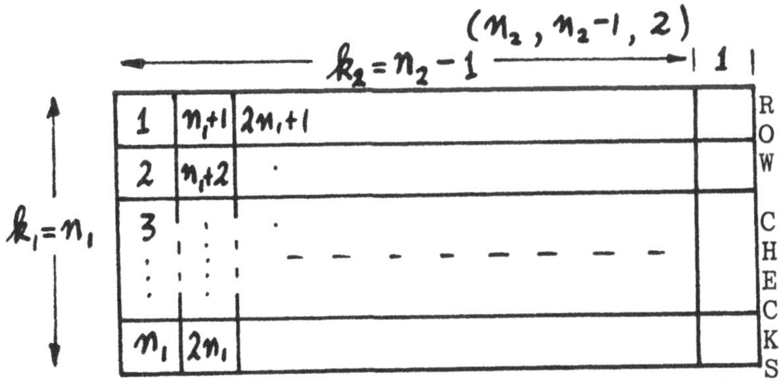

FIGURE 6 : BURST-DETECTING ARRAY CODE ($b=n_1$)

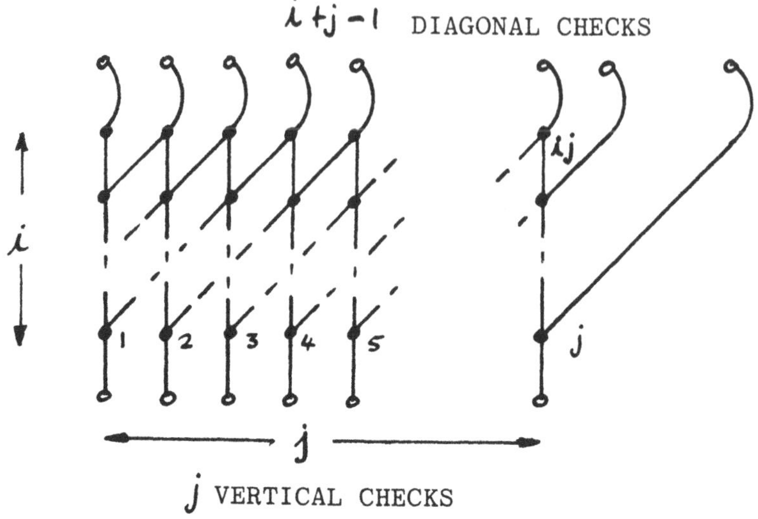

FIGURE 7 : ORTHOGONAL CHECKS FOR BURST CORRECTION

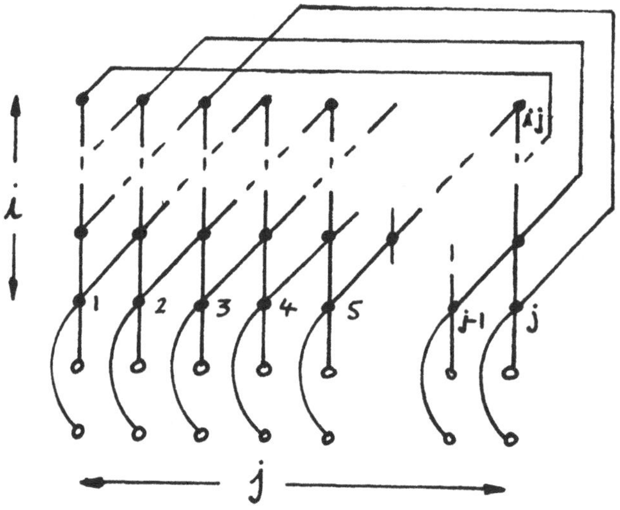

FIGURE 8 : RBC ARRAY CODE

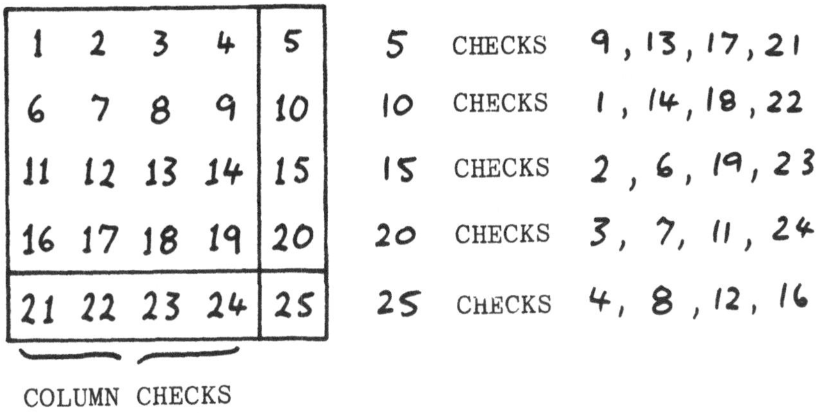

FIGURE 9 : SQUARE (5x5) RBC CODE

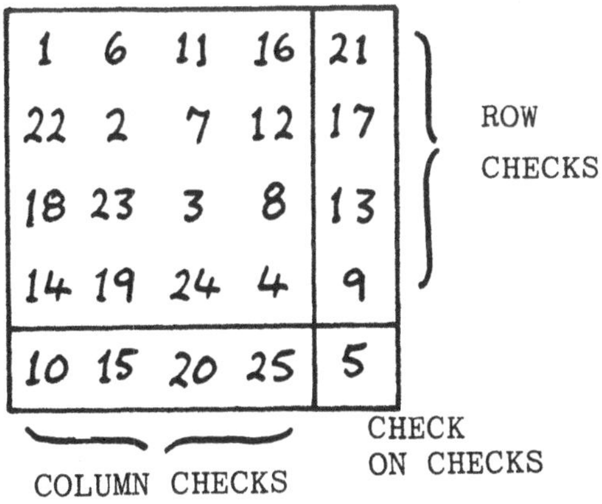

FIGURE 10 : SQUARE ARRAY CODE WITH DIAGONAL READ-OUT

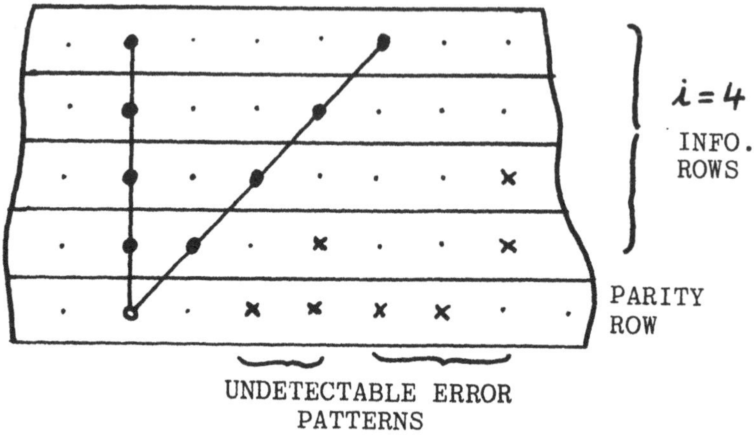

FIGURE 11 : A MOVING PARITY (CONVOLUTIONAL) ARRAY CODE

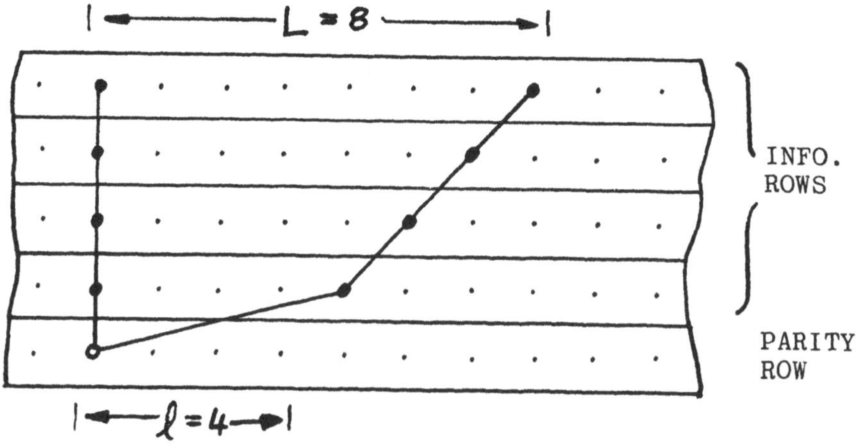

FIGURE 12 : A MOVING PARITY RBC CODE

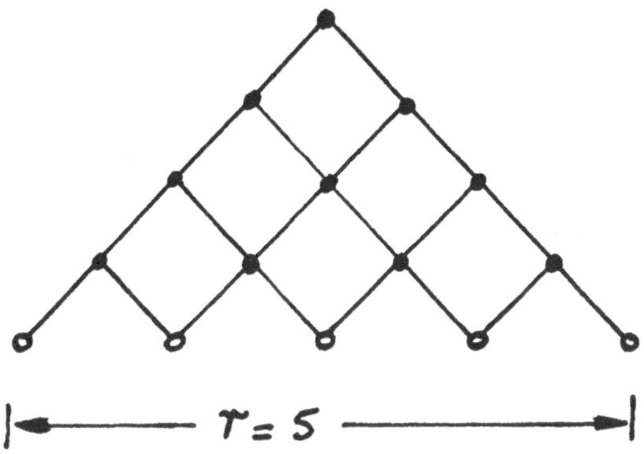

FIGURE 13 : A (15,10,3) WING CODE

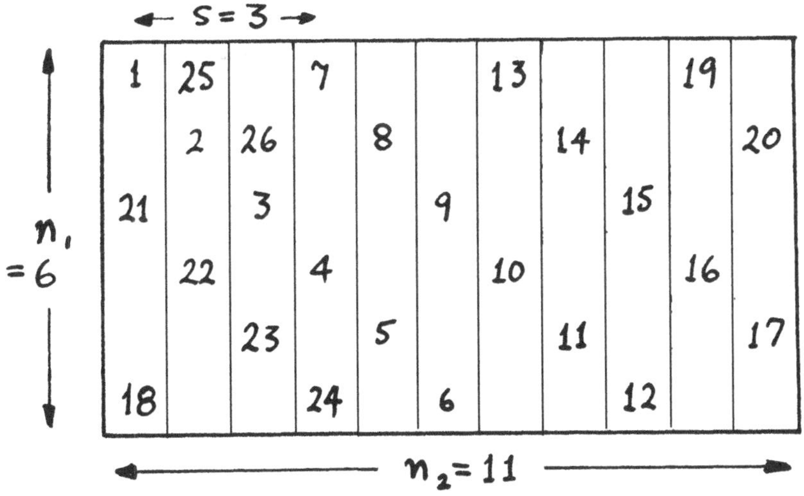

FIGURE 14 : AN SBC ARRAY CODE WITH DIAGONAL READ-OUT PARAMETER s=3

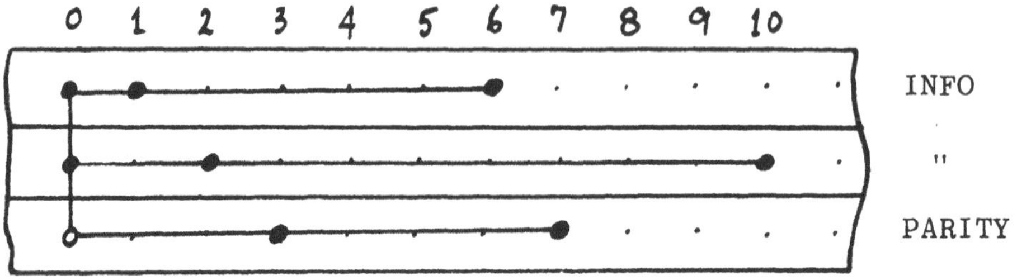

FIGURE 15 : A RATE 2/3, DOUBLE-ERROR-CORRECTING ORCHARD CODE

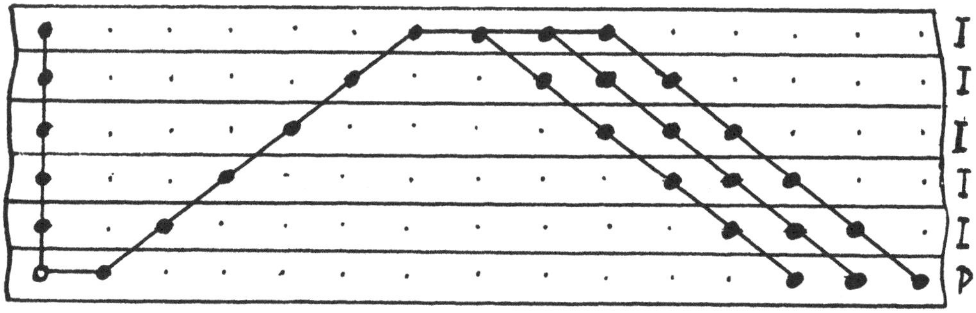

FIGURE 16 : A RATE 5/6, DOUBLE-ERROR-CORRECTING, TRIPLE-ERROR-DETECTING ORCHARD CODE

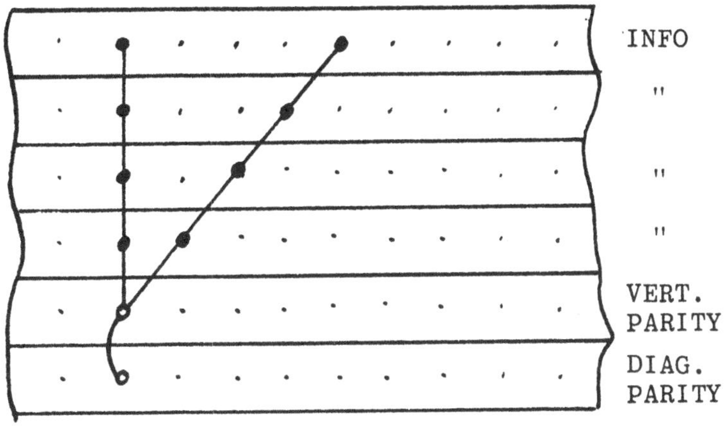

FIGURE 17 : A SIMPLE CROSS-PARITY CODE

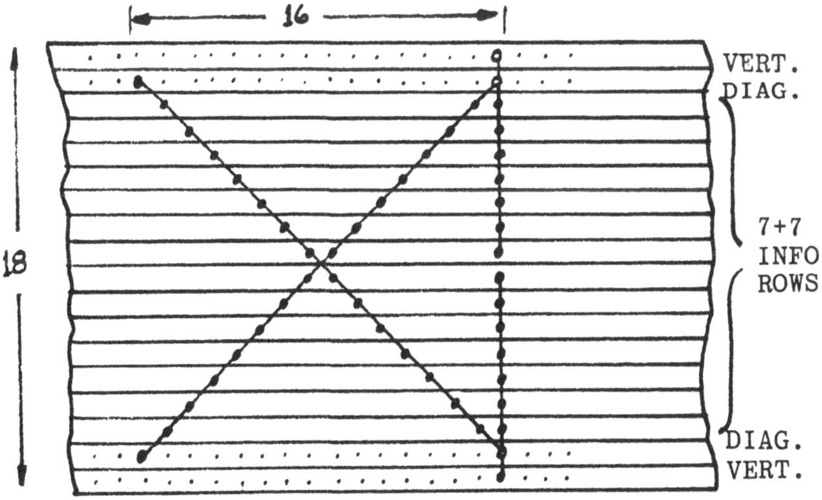

FIGURE 18 : ADAPTIVE CROSS-PARITY CODE

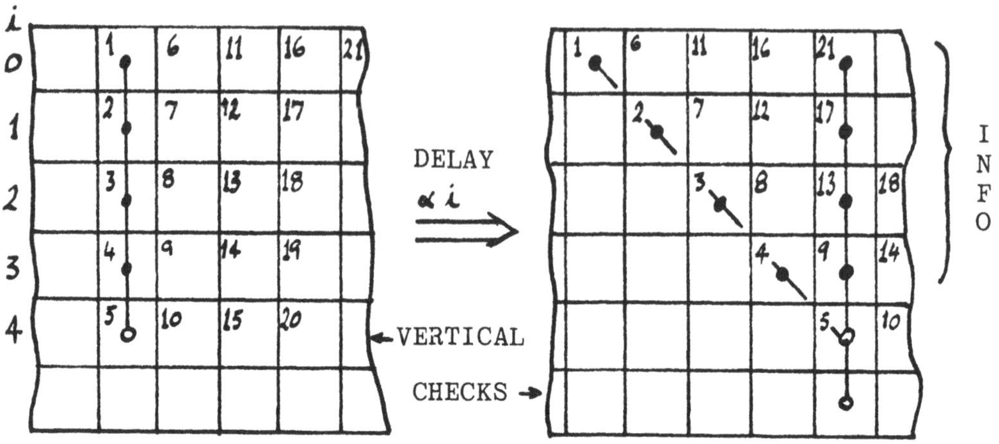

FIGURE 19 : CROSS-INTERLEAVED CODE

PART TWO

Geometries and Cryptography

AN INTRODUCTION TO THE THEORY OF UNCONDITIONAL SECRECY AND AUTHENTICATION

A. Sgarro
Università di Udine, Udine, Italy
and
Università di Trieste, Trieste, Italy

Abstract

A few insecure ciphers, a seemingly secure cipher and a provably secure ciphers are described to illustrate the basic points in Shannon-theoretic cryptography. Lower bounds for authentication codes in Simmons' authentication theory are given.

PART ONE: SECRECY

1 Cipher systems

Below, the coat-of-arms of (*classical, symmetric, secret-key*) cryptography is described.

A (*clear*)*message* x is *enciphered* (*encrypted*) to provide a *cryptogram* y, y is transmitted over a noiseless channel and *deciphered* at the sink by the legitimate receiver. The legal operations of enciphering and deciphering are carried out using a piece of top-secret information z, called the *key*; the key is transmitted to the receiver via a secure (but "costly") channel, which, unlike the cryptogram channel is not subject to tapping. The cryptogram y is intercepted by the opposer (the cryptanalyst) who tries to *decrypt* it (to decipher it without knowledge of the key).

In a good cipher system, deciphering is very easy and decrypting is very hard, maybe even impossible. We accept *Kerckhoffs' principle* ruling strategic (long-term) cryptography: the soundness of a cipher system does not (essentially) depend on keeping secret the type of cipher system being used, but only on keeping secret the particular key z being used. Consequently, we shall always assume that the opposer knows for free which type of cipher system is being used.

Notice that in classical cryptography two channels are there: this is not the case in *public-key* (*asymmetric*) cryptography, a brand-new approach born in the late 70's. In this case the enciphering key is public information, and only the deciphering key (which, however, has no need to circulate) is kept secret; splitting the key in this way is a bright idea which has been made possible by progress in complexity theory. Only symmetric cryptography lies within the scope of this contribution.

More formally, the elements of a cipher system are a finite random triple XYZ (random message, cryptogram and key) over a product alphabet $\mathcal{X} \times \mathcal{Y} \times \mathcal{Z}$ as follows:

i) X and Z are independent (correlation might help the spy)

ii) X is a deterministic function of YZ: $X = g(YZ)$

(To avoid trivial specifications, we rule out messages, cryptograms or keys whose (marginal) probability is zero). We shall also assume, unless otherwise specified:

iii) Y is a deterministic function of XZ: $Y = f(XZ)$

iv) the cipher system is endomorphic (the alphabets of X and Y, $\mathcal{X}$ and $\mathcal{Y}$, coincide)

v) the cipher is canonical (the distribution of Z is uniform, so as to avoid probable keys)

Functions f and g mentioned in iii) and ii) correspond to enciphering and (legal) deciphering, respectively; for $Z = z$, f and g are inverse of each other. We warn the reader that for some cipher systems in use iii) is violated.

Because of i) to iii) if cryptogram y corresponds to message x under key z one has:

$$\Pr(Y = y, Z = z) = \Pr(X = x)\Pr(Z = z) \tag{1}$$

The safety of a cipher system can be assessed by computing relevant success probabilities. The opposer can play three "games": key estimation, message estimation (= decryption) and key-appearance estimation.

Let ϕ, ψ and ξ be the strategies adopted by the opposer to play these games, that is to estimate the key or the message given the cryptogram, and to estimate the key given the couple message-cryptogram, respectively. Formally, ϕ, ψ and ξ are three deterministic functions, $\phi : \mathcal{Y} \to \mathcal{Z}$, $\psi : \mathcal{Y} \to$

$\mathcal{X}$, and $\xi : \mathcal{X} \times \mathcal{Y} \to \mathcal{Z}$. The last is useful in case of so-called *plaintext attacks*, when the same key will be used for encryption also in the future; a plaintext attack is feasible when, e.g., the spy gets hold of the enciphering equipment for a while so as to encrypt fake messages and observe the output cryptograms. We shall append the subscript n when we want to emphasize the length of the cryptogram.

If one wants to estimate the value taken by random variable A after observing the value b taken by random variable B, the best thing he can do is to choose a such as to maximize the success probability $\Pr(A = a \mid B = b)$. (Statisticians will observe that we are adopting Bayesian estimation). Then, the overall success probability of an optimal strategy is

$$\sum_b \Pr(B = b) \max_a \Pr(A = a \mid B = b) = \sum \max \Pr(A = a, B = b)$$

$$\geq \max_a \sum_b \Pr(A = a, B = b) = \max_a \Pr(A = a) \qquad (2)$$

The lower bound is attained whenever A and B are independent (independence is not a necessary condition for equality); its value is the best probability of successful estimation of A when no side information B is available.

We shall assume that the spy always adopts *optimal* strategies ϕ, ψ and ξ. The corresponding optimal success probabilities will be denoted by P_K, P_M, P_{KApp}, respectively (the argument n will be added when needed to emphasize the length of the cryptogram).

Given the cryptogram the correct key gives back the correct message: then each strategy for estimating the key gives automatically a strategy for estimating the message (for decryption) which has at least the same probability of success. Then

$$P_M \geq P_K$$

(This fact has also another consequence: one might be interested in a fourth type of attack, the one against the couple key-message given the cryptogram; however, our observation implies that this attack reduces to key estimation given the cryptogram).

Observe that by combining the optimal message estimation strategy and the optimal key-appearance estimation strategy we obtain a two-step key estimation strategy. If the partial strategies are successful, so is the two-step strategy; this gives a lower bound for its success probability, and a fortiori for P_K:

$$P_K \geq \Pr(\psi \text{ succeeds and } \xi \text{ succeeds})$$

By double negation, De Morgan's rule and the union bound:

$$P_K \geq 1 - \Pr(\psi \text{ fails}) - \Pr(\xi \text{ fails}) = P_M - (1 - P_{K\,App})$$

Then:

$$P_K \leq P_M \leq P_K + (1 - P_{K\,App}) \tag{3}$$

By streamlining the definitions given in /1/ we define the cipher system XYZ to be:

ideal if Y and Z are independent

perfect if X and Y are independent

(the cryptogram by itself does not leak any information about the key, or the message, respectively, since the opposer can do as well by simply ignoring the intercepted cryptogram; cf bound (2)).

We observe that even if the success probabilities are small a cipher system might be secure in practice: this happens when the computational effort to recover Z, or X, out of Y is unbearable, so as to make the estimation feasable only in principle, but not in real life. Below we deal with *unconditional security*, that is we assume that the spy has unlimited computational capacity. The idea of being contented with *computational security* is made good use of in public-key (asymmetric) cryptography.

2 A few insecure ciphers

For the sake of illustration we begin by putting forward the *additive* cipher, or Caesar's cipher. Here $\mathcal{X} = \mathcal{Y} = \mathcal{A}^n$, $\mathcal{A}$, the single-letter alphabet, being the additive group of residues modulo m. (For $m = 26$ we can agree that 0 stands for A, 1 for B, ..., 25 for Z). One has $y_i = x_i + z \bmod m$, $y^n = y_1 y_2 \ldots y_n$, $x^n = x_1 x_2 \ldots x_n$ (we add the superscript n to emphasize that messages and cryptograms are n-length sequences). Breaking this cipher is child's play: the key space being quite small, *exhaustive search* will do. A good (strategic) cipher has to have a large key space: this condition, however, is far from sufficient to ensure secrecy, as we show below.

Adding z to the elements of $\mathcal{A}$ gives a *permutation* of $\mathcal{A}$. We shall allow that z be *any* permutation of $\mathcal{A}$, so as to have $m!$ keys (the algebraic structure of $\mathcal{A}$ is no longer needed). For $m = 26$, $m!$ is a pretty large number, or at least it was before computers came into use. The resulting cipher is called *simple substitution cipher*: one has $y_i = z(x_i)$, z being a permutation of the single-letter alphabet; $x_i = z^{-1}(y_i)$. Actually, such a cipher can be broken by a statistical analysis, as masterly explained by E. A. Poe in his tale *The golden bug*, and as known to the Arabs around 1000 A.D.: the reason is that ciphertext letters "inherit" the characteristic frequencies of the corresponding cleartext letters.

Below we investigate this fact with more care in the case when the message source is *stationary* and *memoryless* (the random cleartext letters $X_1, X_2, \ldots, X_n$ which make up the random message X^n are independent and have the same probability distribution $P = (p_1, p_2, \ldots, p_m)$ over $\mathcal{A} = (a_1, a_2, \ldots, a_m)$, $p_j > 0$, $\sum p_j = 1$).

We shall need the following unpretentious form of the (weak) law of large numbers: for each n there is a subset $\mathcal{T}_n$ in $\mathcal{A}^n$ such that: $P^n(\mathcal{T}_n) \to 1$; relative frequencies f_j with which each letter a_j occurs in the sequences of $\mathcal{T}_n$ belong to intervals $(p_j \pm \delta_n)$ with $\delta_n \to 0$. Sequences in $\mathcal{T}_n$ will be called *typical sequences*.

Now assume that we are able to find out how much the probability of an unknown letter in $\mathcal{A}$ is: this piece of information is enough to identify that letter iff no other letter in $\mathcal{A}$ has the same probability; letters with the same probability cannot be discriminated among.

We declare equivalent two keys when, with respect to each other, they only garble letters which have the same probability; more precisely: $z \sim \zeta$ iff $P(z^{-1}(a_j)) = P(\zeta^{-1}(a_j))$, $1 \leq j \leq m$. Assume that the distinct components in P are s $(1 \leq s \leq m)$, and that they occur $t_1, t_2, \ldots, t_s$ times in P, respectively $(t_1 + t_2 + \ldots + t_s = m)$. The number of keys in each equivalence class is clearly: $\alpha = \alpha(P) = t_1!t_2!\ldots t_s!$ $(1 \leq \alpha \leq m!$; $\alpha = 1$ iff the components of P are all distinct, $\alpha = m!$ iff they are all equal).

Now, if n is large enough so as the s intervals centered in each distinct p_j and of width $2\delta_n$ are disjoint the spy is able to identify correct probabilities whenever the clearmessage is typical; that is, he can devise a strategy $\omega_n = \omega_n(Y^n)$ mapping the set of cryptograms over the set of key-classes such that $\Pr(W = \omega_n(Y^n))$ goes to one with n ; W is the random key-class to which key Z belongs: $W = w(Z)$. Notice that ω_n is not a key estimation strategy, but only a key-class estimation strategy, so the spy's job is not over, unless all the probabilities in P are distinct.

A simple check (use (1)) shows that, given Y^n and W, Z is uniform over its class:

$$\Pr(Z = z \mid W = w, Y^n = y_n) = \alpha^{-1}$$

if $z \in w$, else zero. (As one has also $\Pr(Z = z \mid W = w) = \alpha^{-1}$, Z is independent of Y^n given W). For uniform distributions, the optimal strategy is simply haphazard guessing, which here gives success probability equal to $\frac{1}{\alpha}$.

We have devised a two-step strategy ϕ'_n to estimate the key; its error probability has the form:

$$\varepsilon_n + (1-\varepsilon_n)\frac{\alpha-1}{\alpha} = \frac{\alpha-1}{\alpha} + \epsilon_n$$

with $\varepsilon_n, \epsilon_n \to 0$ (either one fails in the first step, and then the message is not typical, or, assuming the first step is successful, one fails in the second step).

If $\approx$ means "equal up to a quantity which vanishes with n", one has for the success probability relative to the two-step strategy: $P'_K(n) \approx \frac{1}{\alpha}$

Actually, this asymptotic result cannot be improved upon, as follows from the (non-asymptotic) inequality $P_K(n) \leq \frac{1}{\alpha}$ (were this false, one would have a fortiori, by simply ignoring the key-class, a second-step strategy with success probability better than haphazard guessing, which is absurd).

It will be shown below that $P_{KApp}(n)$ goes to 1. Then (use (3)):

$$P_K(n) \approx P_M(n) \approx \frac{1}{\alpha}$$

We go now to the estimation of Z given X^nY^n. Observe that, whenever at most one letter is missing in x^n (or, equivalently, in y^n) z can be correctly estimated from the couple x^ny^n. This implies that, if n is large enough so as the interval centered in the lowest probability in P and of width $2\delta_n$ does not cover zero, there exists a strategy which is successful whenever X^n is typical (for n large enough no letter can be missing in a typical sequence). Then:

$$P_{KApp}(n) \approx 1$$

We have proved the following:

i) key and message are successfully estimated with probability which goes to 1 unless there is at least a tie in P

ii) for P uniform a simple substitution cipher is ideal

Ruling out the case rarely met in practice that n is small, it appears that a substitution cipher is safe to use when the following conditions are both met: cleartext attacks are unfeasible and there are many ties in P.

In practice, simple substitution ciphers are known to be unsafe. Actually, even if cleartext attacks are unfeasible, a small error in the evaluation of the p_j's might break up the ties in P with catastrophic consequences (in the statistical terminology, our asymptotic results are not robust). A more substantial objection will be put forward in the next section.

3 A seemingly secure cipher

In a *transposition cipher* the cleartext is chunked into blocks of length T ($T \geq 2$ is called the *period* of the cipher) and each block is shuffled (anagrammed) according to the key, which is a fixed permutation of the the integers from 1 to T. In the past this shuffling was often done using very nice revolving grills, as depicted in the novel *Mathias Sandorf* by Jules Verne. If all keys are made use of, as we assume below, a transposition cipher has $T!$ keys. We also assume that the cleartext is stationary and memoryless; the message length n will be constrained to be a multiple of T.

Using (1) one easily finds that Z and Y^n are independent; we shall show below that $P_{K\,App}(n)$ goes to one; then (cf (3)):

$$P_M(n) \approx P_K(n) = \frac{1}{T!}$$

(the second equality holds also for finite lengths, while the first is only asymptotic).

We go to the computation of $P_{K\,App}(n)$. First we define the *index of coincidence* $C(P) = \sum p_j^2$ (called also other names: informational energy, or, less pretentiously, square of the Euclidean norm). Developing $(\sum p_j)^2$ and $\sum(p_j - \frac{1}{m})^2$ one finds: $\frac{1}{m} \leq C(P) \leq 1$, with equality on the left iff P is uniform, on the right iff P is deterministic; in our case $C(P) < 1$.

Z can be deterministically computed from the couple X^nY^n unless the following fact occurs: in each block (message-block, say; cryptogram blocks would also work) there are at least two positions u and v such that $x_{(r-1)T+u} = x_{(r-1)T+v}$, $1 \leq u, v \leq T$, $1 \leq r \leq n/T$. The probability of this unpleasant event is upperbounded by $\binom{T}{2}C(P)^{n/T}$ (first fix u and v and then use the union bound), and the upper bound vanishes with n. then:

$$P_{K\,App}(n) \approx 1$$

We have proved the following:

i) key and message are successfully estimated with probability bounded away from 1

ii) a transposition cipher is ideal

Now, unlike in the case of simple substitution ciphers, claim ii) is robust, and is not affected by changes in the p_j's. It would appear, therefore, that a long-period transposition cipher is safe to use whenever cleartext attacks are unfeasible. In practice, transposition ciphers are not safe to use, as century-long experience shows. The reason is the following. The results of sections 2 and 3 may well be mathematically nice, but they hold only for idealized stationary and memoryless message sources. These sources are nowhere to be found in real life; consequently, the cryptanalyst may avail himself of a lot of correlation in the cleartext, e.g. couple frequencies to break transposition ciphers, or even long-range correlation which is not to be squeezed into the fetters of mathematically clean statistical models (of the type: just before turning page 1000 of the feuilleton I'm reading I foresee that the first letter at page 1001 will be an A, because the last word at page 1000 was Aleksej, and it is high time that Aleksej Andrejevič, last mentioned at page 10, shows up to the rescue of the unhappy orphan girl).

4 A provably secure cipher

We are in trouble! Let's explore the ways out.

Ciphers which by themselves are weak can be combined to give a (hopefully) secure *composite* (*cascade*) cipher. E.g. DES, or Data Encryption Standard, put forward by IBM and adopted in 1977 by the US National Bureau of Standards for protection of sensitive data (but not top-secrets involving national security), makes use of iterated substitutions and transpositions. The composition has been devised very well indeed, since so far DES is unbroken even if its algorithm is public knowledge (actually, DES is the first cipher in serious use whose algorithm has been made public: by so doing, for the first time Kerckhoffs' principle has been led to its extreme consequences, which is quite telling of the new sociology of commercial cryptography brought about by large computer networks).

The key-space of DES has size 2^{56}, which, however, seems to be becoming a small number, so that the next version of DES will have to have a much larger key-space! Actually, it is not even known whether DES is *faithful*, so that the number of distinct key transformations might be smaller than 2^{56}. (In an unfaithful cipher there are at least two keys z and ζ such that $g(z,y) = g(\zeta,y)$ whatever the cryptogram y. In this case the term *key-label* seems more appropriate; one might instead reserve the more committal term "key" for equivalence classes of key-labels which decipher in the same way. Note that P_K is a significant cryptographic measure of safety only when Z is such a class of key-labels. Cf /2/ for a discussion in the case of homophonic ciphers. All the ciphers covered here are faithful).

Let us pursue another way out. First we go back to additive ciphers and also to their Renaissance users who possessed rotable discs to implement them. An idea to improve the performance of the cipher (to flatten the statistics of the ciphertext) is to make use of a keyword, or of a key-sentence, rather than of a key-letter: if $z_1 z_2 \dots z_T$ is the key, $y_i = x_i \oplus z_r$, where r is the remainder when dividing i by T (set $r = T$ when the remainder is zero). The resulting cipher is called *Vigenère cipher*; T is the period of the cipher. Vigenère cipher is the simplest example of a *polyalphabetic* cipher (several substitutions are used in turn). The earliest polyalphabetic ciphers were proposed by Leon Battista Alberti, the famous humanist scholar, in the

15th century: however, polyalphabetic ciphers were first broken only one century ago, when Fr. von Kasiski noticed that repetitions in the ciphertext occur at a distance that is likely to be a multiple of T : once T is found out, the solution of a polyalphabetic cryptogram is reduced to the solution of T monoalphabetic sub-cryptograms.

It soon occurred to cryptographers that to strenghten a polyalphabetic cipher it was better to use long and weird keys. The *one-time pad*, the cipher we are going to investigate, is nothing else but the asymptotic (non-periodic) and randomized version of the Vigenère cipher. More specifically, we assume that the key-source is *totally random* (stationary, memoryless, Z_i uniform: e.g., for $m = 2$, a coin-flipping process); set $y_i = x_i \oplus z_i$ (of course the key generator operates independently of the message source). We emphasize that in the following we are making *no* assumption on the statistics of the clearmessage source: there might be intractable correlation, like the one with Aleksej Andrejevič.

First, we prove that also the cryptogram source $Y_1 Y_2 \dots Y_n \dots$ is totally random. Actually, by (1), $\Pr(Y^n = y^n, Z^n = z^n) = \frac{1}{m^n} \Pr(X^n = y^n - z^n)$. Summing over z^n, $y^n - z^n$ spans $\mathcal{A}^n$, and so $\Pr(Y^n = y^n) = \frac{1}{m^n}$.

Second, we prove that message and cryptogram are independent, that is that the one-time pad is a perfect cipher. One has:

$$\Pr(X^n = x^n, Y^n = y^n) = \Pr(X^n = x^n)\Pr(Z^n = y^n - x^n)$$

$$= \Pr(X^n = x^n)\Pr(Y^n = y^n)$$

We now argue that the one-time pad is perfect not only in the technical, or formal, sense of the world: actually, perfection is *universal* (holds independently of the message statistics); moreover, Z_n^h is obviously independent of X^n, and even of X^nY^n, so that also attacks on the (future) key are unfeasible ($Z_n^h = Z_{n+1}Z_{n+2}\ldots Z_h$, $h > n$; of course Z^n is deterministically computable out of X^nY^n, but Z^n is of no use to the cryptanalyst).

If the one-time pad is used, the best strategy for the spy is guessing without even looking at the intercepted cryptogram! The draw-back of the one-time pad is that the key is at least as long as the message: this makes it unusable in most practical situations. However, this drawback is unremovable.

Actually, in a perfect cipher, $\Pr(X = x, Y = y) = \Pr(X = x)\Pr(Y = y)$ and so is positive for any couple xy; if y is kept fixed, to obtain all the messages x by deciphering y one needs as many distinct keys z as there are distinct messages x: then $|\mathcal{Z}| \geq |\mathcal{X}|$. In the case of the one-time pad $|\mathcal{Z}| = |\mathcal{X}|$, which is the lowest possible value: in this sense the one-time pad is the perfect cipher with the less key material. For a different, and possibly deeper, approach to the optimality of the one-time pad as a perfect cipher cf the appendix.

In practice one makes use of pseudo-random key letters, and so one has to be contented with "pseudo-perfection", rather than genuine perfection. The question is: how well does pseudo-perfection approximate genuine perfection?

Unfortunately, if one makes use of standard pseudo-random number generators, the results are catastrophic! (Cf e.g. /3/). The problem of finding special generators which are suitable for cryptographic use is one of the main topics of research in modern cryptology.

Notice that, to obtain a totally random, or "slippery", cryptogram, rather than randomizing the key, one might try to randomize the message source by suitably encoding it. Were the encoded message totally random, any additive cipher system with sufficiently many keys would do, because the cipher system would necessarily be ideal! (We are interchanging key and message with respect to the one-time pad.) This is one of the ideas behind *homophonic* ciphers; in this case enciphering is *not* deterministic (deciphering of course is). After remaining dormant for nearly two centuries, homophonic enciphering has been again paid attention to (cf /2/, /4/, /5/), and might turn out to be a promising tool in future cryptologic research aimed at unconditional security.

In Shannon-theoretic cryptography one usually assesses the strength of cipher systems by computing relevant equivocations which measure the uncertainty of the spy, rather than by computing relevant probabilities of successful estimation, as we have done. However, the approach based on optimal strategies which we have used here is conceptually, even if not necessarily analytically, simpler. Rather deep results for substitution and transposition ciphers are to be found in /6/, /7/, /8/, /9/ in the case of "idealized" (stationary and memoryless) message sources; in particular, a more careful analysis of convergence speed is done that the one offered here, and finite-length bounds are provided. Cf also /10/; for a comparison of the two approaches cf /11/.

In the case of authentication coding, as covered below, the approach based on optimal strategies is quite standard.

5 Appendix: the equivocation approach

In this appendix we assume some familiarity with information measures, as can be derived, e.g., from the first chapters of /12/.

Following the seminal paper /1/ the safety of a cipher can be assessed by computing three relevant equivocations (conditional entropies), $H(X \mid Y)$, $H(Z \mid Y)$ and $H(Z \mid XY)$ rather than three relevant success probabilities. These equivocations, which are called *message equivocation, key equivocation* and *key-appearance equivocation*, respectively, represent posterior uncertainties after interception of Y (of XY in the case of plaintext attacks).

By developing in two ways $H(XZ \mid Y)$ and by recalling that deciphering is deterministic ($H(X \mid YZ) = 0$) one finds that the three equivocations are functionally constrained to one another in a very simple way (cf instead (3)):

$$H(Z \mid Y) = H(X \mid Y) + H(Z \mid XY)$$

and, in particular:

$$H(Z \mid Y) \geq H(X \mid Y)$$

The counterpart to (1), which assumes also deterministic enciphering, is instead:

$$H(YZ) = H(X) + H(Z)$$

A cipher is ideal when $H(Z \mid Y) = H(Z)$, perfect when $H(X \mid Y) =$

$H(X)$ (prior and posterior equivocations coincide; it is well-known that this happens precisely under independence).

The results of sections 2 to 4 can be easily recycled to the case of equivocations using the following continuity property which is an obvious consequence of Fano's inequality: if U and V_n are correlated random variables and if there exists a deterministic function of V_n, $f_n = f_n(V_n)$, such that $\Pr(U \neq f_n(V_n))$ vanishes with n, then also the equivocation $H(U \mid V_n)$ vanishes with n. ($f_n(V_n)$ is interpreted as an estimation of U formed upon observing V_n; one requires that the error probability goes to zero; Fano's inequality implies directly $H(U \mid f_n(V_n)) \to 0$, however $H(U \mid V_n) \leq H(U \mid f_n(V_n))$).

In the case of simple substitution ciphers one also needs the equality

$$H(Z \mid Y^n) = H(W \mid Y^n) + H(Z \mid WY^n)$$

obtained by developing in two ways $H(ZW \mid Y^n)$ and by recalling that the random key-class W is a deterministic function of the random key Z. Below we also use the obvious fact that the entropy of a uniform random variable is log the number of its values. The recycled results follow below.

Simple substitution ciphers:

$$H(Z \mid Y^n) \approx H(X^n \mid Y^n) \approx \log(\alpha)$$

$$H(Z \mid X^n Y^n) \approx 0$$

Transposition ciphers:

$$H(Z \mid Y^n) = \log(T!) \approx H(X^n \mid Y^n)$$

$$H(Z \mid X^n Y^n) \approx 0$$

One-time pad:

$$H(X^n \mid Y^n) = H(X^n)$$

$$H(Z_n^h \mid Y^n) = H(Z_n^h \mid X^n Y^n) = H(Z_n^h)$$

We are now ready to give the deep proof of the optimality of the one-time pad as a universal perfect cipher, as promised in section 4.

Under perfection $H(X \mid Y) = H(X)$ and so one has:

$$H(Z) \geq H(Z \mid Y) \geq H(X \mid Y) = H(X)$$

If perfection is universal this holds whatever the distribution of X, and so:

$$H(Z) \geq \log(|\mathcal{X}|)$$

In the case of the one-time pad the lower bounds holds with equality. This is what we need since entropy can be identified with shortest encoding length as implied by Shannon's fundamental source-coding theorem.

PART TWO: AUTHENTICATION

6 An introductory remark

Formally, authentication codes are again random triples XYZ for which i) and ii) as in section 1 hold. However, as we shall see below, the purpose to which they are appointed is different. Often also iii) is required; instead (cf below) requiring iv) would be catastrophic.

The following sections are an almost literal quotation of /13/, save an immaterial omission and an addition on combinatorial bounds. In sections 8 and 9 we shall deal with "abstract" codes YZ: by this we mean that the couple YZ is absolutely general, and not necessarily derived from an authentication code XYZ. General references to authentication are e.g. /14/ or /3/. Note that often, as in /14/ and /15/, the cryptographic-flavoured terms clearmessage, cryptogram and key as used here are replaced by the terms source-state, message (helas!) and encoding rule, respectively; a third alternative, more in keeping with the usage of non-secret coding theory, would be: message, codeword and encoding rule.

7 Preliminaries

Below we shall put forward a basic lower bound in authentication theory which is due to Simmons (cf e.g. /14/ or /3/); as observed in /13/ this bound is a straightforward consequence of the non-negativity of the *infor-*

mational divergence:

$$D(P,Q) = \sum_i p_i \log \frac{p_i}{q_i} \geq 0 \ , \ D(P,Q) = 0 \text{ iff } P = Q \qquad (4)$$

($D(P,Q)$ is called also: discrimination, cross entropy, Kullback-Leibler's number; above P and Q are two probability distributions with the same number of components; logs are to a base greater than 1, e.g. 2 or e; for $a > 0$ set $0 \log \frac{0}{a} = 0 \log \frac{0}{0} = 0$, $a \log \frac{a}{0} = +\infty$). The proof of (4) is itself quite short: one has $\ln x \leq x - 1$ with equality iff $x = 1$ (ln is strictly concave and lies below its tangent at $x = 1$); set $x = \frac{q_i}{p_i}$, multiply by -1 and plug the result into the definition of $D(P,Q)$.

A remarkable divergence is the *mutual information* $I(Y,Z)$; here P is the joint distribution of the random couple YZ and Q is the joint distribution obtained by multiplication of the marginal probabilities "as if" Y and Z were independent; (4) implies that the mutual information is non-negative, and is zero iff Y and Z are independent.

Actually, we need conditional divergences: if YZ and YW are two random couples over the same product-space the *conditional divergence* $D(Z,W \mid Y)$ is defined as the weighted average $\sum_y \Pr(Y = y) D(P_y, Q_y)$, where P_y (Q_y, respectively) is the conditional distribution of Z (of W, respectively) given $Y = y$. Of course, by (4):

$$D(Z,W \mid Y) \geq 0 \text{ , with equality iff } YZ \simeq YW \qquad (5)$$

("$\simeq$" denotes equality of distributions)

Below we review the easy derivation of Simmons' bound and show that it can be extended to substitution games.

8 Impersonation games

Let YZ be the random couple cryptogram-key. In the impersonation game the opponent chooses a cryptogram y hoping it to be taken as legal. The probability of success is

$$\Pr(Z \in A_y) \tag{6}$$

$$\text{with } A_y = \{z : \Pr(Y = y, Z = z) \neq 0\} \tag{7}$$

(A_y is the set of keys which authenticate y). The best strategy for the opponent is to choose y as to maximize the probability in (6). Let P_I be this maximal probability of success in the impersonation game:

$$P_I = \max_y \Pr(Z \in A_y)$$

Set $\Pr(W = z \mid Y = y) = \Pr(Z = z \mid Z \in A_y)$. Plug this into (5) and recall the definition of the mutual information $I(Y, Z)$ to obtain

$$E_Y \log \Pr(Z \in A_Y) \geq -I(Y, Z)$$

(E_Y denotes expectation.) Noting that P_I is the maximal value for the random probability $\Pr(Z \in A_Y)$ one has Simmons' bound (we take binary logs):

$$P_I \geq 2^{-I(Y,Z)} \tag{8}$$

The conditions for equality are soon found: YZ and YW must have the same distribution, and $\Pr(Z \in A_y)$ must be constant in y. An authentication system for which (8) holds with equality will be called, in the spirit of Simmons' definition, *perfect* (actually, perfection as defined by Simmons is more ambitious and covers also the substitution game: ours is perfection with respect to the impersonation game only).

We note that the original derivation of (8) is 'tedious and long', and even the streamlined derivation given in /3/ is 'somewhat lengthy' (we are quoting /14/ and /3/, respectively).

9 Substitution games

Assume cryptogram c has been legally sent (and assume that there are at least two cryptograms). The opponent deletes c and sends instead $y \neq c$; his probability of success is now:

$$\Pr(Z \in A_y \mid Y = c) = \Pr(Z \in A_y \cap A_c \mid Y = c) \tag{9}$$

The best strategy is to maximize (9) with respect to $y \neq c$. Let $P_S(c)$ be the corresponding maximal probability of success:

$$P_S(c) = \max_{y \neq c} \Pr(Z \in A_y \cap A_c \mid Y = c)$$

Set $a_c(z) = \sum_{y \neq c} \Pr(Y = y \mid Z = z)$. $a_c(z) = 0$ means that z authenticates *only* c (z is a "one-cryptogram key"). Assume that $a_c(z) \neq 0$ for all z in A_c, and define the random couple $Y_c Z_c$ as follows:

$$\Pr(Z_c = z) = \Pr(Z = z \mid Y = c) \quad , z \in A_c \tag{10}$$

$$\Pr(Y_c = y \mid Z_c = z) = a_c(z)^{-1} \Pr(Y = y \mid Z = z) \quad , y \neq c \tag{11}$$

Note that in $Y_c Z_c$ the conditional probabilities of the cryptograms have been "pumped up" so as to sum to 1 (c is no more there). Let us play the impersonation game for Y_cZ_c: one has (8) with Y_cZ_c instead of YZ. If $P_I(c)$ denotes the probability of success for the ancillary impersonation game, one has:

$$P_I(c) = \max_{y \neq c} \Pr(Z_c \in A_y^c) = \max_{y \neq c} \Pr(Z \in A_y^c \mid Y = c)$$

$$= \max_{y \neq c} \Pr(Z \in A_y^c \cap A_c \mid Y = c)$$

where A_y^c is defined as in (7) with Y_cZ_c instead of YZ. However, as one easily checks, the set of keys which authenticate y in the ancillary impersonation game is the same as the set of keys which authenticate y in the original substitution game:

$$A_y^c \cap A_c = A_y \cap A_c$$

(to see this, recall definition (7) and observe that $\Pr(Z_c = z, Y_c = y) \Pr(Z = z, Y = c) \neq 0$ iff $\Pr(Z = z, Y = y) \Pr(Z = z, Y = c) \neq 0$). Recalling the definition of $P_S(c)$, we obtain the reduction formula we had announced:

$$P_I(c) = P_S(c) \tag{12}$$

Therefore, the lower bound for $P_I(c)$ is also a lower bound for $P_S(c)$:

$$P_S(c) \geq 2^{-I(Y_c, Z_c)} \ , \ \ Y_cZ_c \text{ defined as in (10),(11)} \tag{13}$$

By averaging with respect to $\Pr(Y = c)$ one obtains a lower bound for $P_S = \sum_c \Pr(Y = c) P_S(c)$, the overall probability of a successful substitution:

$$P_S \geq \sum_c \Pr(Y = c) \ 2^{-I(Y_c, Z_c)} \tag{14}$$

Equality holds in (13) iff the system Y_cZ_c is perfect, in (14) iff the systems Y_cZ_c are perfect for all cryptograms c's.

(13) and (14) have been derived under the assumption $a_c(z) \neq 0$. We shall show below (section 10) that this assumption always holds in cases of interest. In /13/, for reasons of mathematical completeness, however, also the case when some of the $a_c(z)$ are zero is covered.

10 A toy example

As explained in the introductory remark an authentication code is actually a random triple XYZ rather than a random couple YZ: the term added is the random clearmessage. In the case of substitution games we assume enciphering to be deterministic (we rule out *splitting*); formally, our results hold true also for codes with splitting: however, substitution games are meaningful only when substitution of the legal cryptogram implies substitution of the correct clearmessage, that is only when enciphering is deterministic. A nice way to represent the enciphering and deciphering functions for an authentication code XYZ is to draw up a matrix whose row-headers and column-headers are the keys and the clearmessages, respectively, and whose entries are the corresponding cryptograms. Of course the XYZ model is a special case of the YZ model, and so the lower bounds given above still hold. Actually, since the same cryptogram cannot appear more than once in the same row of the matrix (deciphering is deterministic) one-cryptogram keys are ruled out and so bounds (13) and (14) are usable (we are assuming that the matrix has at least two columns, which is an obvious assumption to make when one plays substitution).

Let us work out a toy example with 2 keys, 2 clearmessages and 3 cryptograms:

*	1	2
1	1	2
2	3	1

We assume equiprobability both for the key and the clearmessage; then $\Pr(Y = 1) = \frac{1}{2}$, $\Pr(Y = 2) = \Pr(Y = 3) = \frac{1}{4}$. Clearly, the (optimal-strategy) probability of successful substitution is $\frac{1}{2}$ if cryptogram 1 is substituted, else is 1. So the overall probability of a successful substitution is $P_S = \frac{3}{4}$. Since this value is returned also by bound (14), as one easily checks, our bound for P_S holds here with equality.

In /15/ four lower bounds are given for P_S (theorems 2.6, 2.9, 2.10, 2.14); however they return values strictly smaller than $\frac{3}{4}$ for our toy example. Computer simulations have substantiated the strength of our bound.

11 Final remark

Formula (12) allows one to reduce substitution games to impersonation games. There is a snag, however. Even if we start by a "complete" authentication code XYZ, the ancillary codes we were able to construct are random couples $Y_c Z_c$ and *not* random triples $X_c Y_c Z_c$. We could convert Simmons' bound for impersonation into a bound for substitution only because the former does not involve the random clearmessage in any way.

To deepen our point let us discuss another case. A well-known combinatorial bound is the following:

$$P_I \geq \frac{m}{h} \tag{15}$$

where m and h are the number of messages and of cryptograms, respectively. Its counterpart for substitution games is:

$$P_S \geq \frac{m-1}{h-1} \tag{16}$$

It would appear that (16) is not derivable from (15) via (12), since (15) involves messages, and so is unusable in the reduction formula (12). Actually, it is enough to obtain a slightly more abstract form of (15) to make the derivation possible, as we now show.

Let YZ be an "abstract" code, not necessarily derived from a random triple XYZ, and let us play the impersonation game for it. One has:

$$hP_I \geq \sum_y \Pr(Z \in A_y)$$

$$= \sum_y \sum_{z \in A_y} \Pr(Z = z) = \sum_z \sum_{y \in A_z} \Pr(Z = z)$$

$$= \sum_z \mid A_z \mid \Pr(Z = z) \geq \min_z \mid A_z \mid$$

where A_z is the set of cryptograms y authenticated by key z. So:

$$P_I \geq \frac{\min_z \mid A_z \mid}{h} \tag{17}$$

which does not involve X and which is more general than (15); (actually, if XYZ is a "genuine" code the numerator is at least m).

However, (17) *can* be used in the reduction formula. Since in the ancillary games Y_cZ_c the number of cryptograms (the number of cryptograms authenticated by key z, respectively) is as in the original game minus one, one obtains, for "genuine" codes XYZ:

$$P_S(c) \geq \frac{m-1}{h-1}$$

which implies (16) by averaging with respect to $\Pr(Y = c)$.

Putting together bounds (8) and (15) it comes out that in an authentication code resistant to impersonation attacks the number of cryptograms must be much larger than the number of messages, and there must be a strong correlation between random key and random cryptogram. (Similar results hold for substitution attacks). Of course these conditions are necessary but not sufficient to ensure that the code is good. Constructive results which yield good codes are based on finite geometries: they will be tackled in further talks given at this school.

References

/1/ C. A. Shannon, *Communication theory of secrecy systems*, Bell Syst. Tech. J., vol.28, Oct. 1949, pp. 656-715

/2/ A. Sgarro, *Equivocations for homophonic ciphers*, in *Advances in cryptology*, ed. by Th.Bet, N.Cot and I.Ingemarsson, Lecture Notes in Computer Science 209, Springer Verlag, 1985, pp.51-61

/3/ J. L. Massey, *An introduction to contemporary cryptology*, Proceedings of the IEEE, May 1988, pp. 533-549

/4/ Ch. G. Günther *A universal algorithm for homophonic coding*, in *Advances in Cryptology - EUROCRYPT 88*, ed. by Ch.G.Günther, Lecture Notes in Computer Science 330, Springer Verlag, 1988, pp.405-414

/5/ H. N. Jendal, Y. J. B. Kuhn, J. L. Massey, *An information-theoretic analysis of Günther's homophonic substitution scheme*, Proceedings of Eurocrypt 89, to be published by Springer Verlag

/6/ R. J. Blom, *Bounds on key equivocation for simple substitution ciphers*, IEEE Trans. Info. Th., vol. IT 25, Jan. 1979, pp. 8-18

/7/ J. G. Dunham, *Bounds on message equivocations for simple substitution ciphers*, IEEE Trans. Info. Th., vol. IT 26, Sept. 1988, pp. 522-527

/8/ A. Sgarro, *Error probabilities for simple substitution ciphers*, IEEE Trans. Info. Th., vol. IT 29, March 1983, pp. 190-198

/9/ A. Sgarro, *Equivocations for transposition ciphers*, Rivista di matematica per le scienze economiche e sociali, Anno 8, Fasc. 2., pp. 107-114

/10/ A. Sgarro, *Simple substitution ciphers*, in *Secure digital communications* ed. by G. Longo, CISM Courses and Lectures N.279, Springer Verlag, 1983, pp. 61-77

/11/ A. Sgarro, *Information-theoretic versus decision-theoretic cryptography*, Elektrotechnik und Maschinenbau (now: Elektrotechnik und Informationstechnik), Heft 12, Dez. 1987, pp. 562-564

/12/ I. Csiszár, J. Körner, *Information theory*, Academic Press, 1981

/13/ A. Sgarro, *Informational divergence bounds for authentication codes*, Proceedings of Eurocrypt 89, in press for Springer Verlag

/14/ G.J. Simmons, *A survey of information authentication*, Proceedings of the IEEE, May 1988, pp. 603-620

/15/ D.R. Stinson, *Some constructions and bounds for authentication codes* Journal of Cryptology, 1,1 (1988) pp. 37-52

APPLICATIONS OF FINITE GEOMETRY TO CRYPTOGRAPHY

A. Beutelspacher
Justus-Liebig University Gießen, F.R.G.

Abstract

In this paper we survey several applications of classical geometric structures to cryptology. Particularly we shall deal with authentication schemes, threshold schemes, network problems and WOM-codes. As geometric counterparts we shall nearly exclusively deal with projective spaces and their internal structures provided by, for example, linear subspaces, Baer subspaces, quadrics, etc.

1. Introduction

1.1 Basic Notions in Cryptology

Cryptology (also called cryptography) provides tools against two types of attacks, namely against the **passive** attack, where the "bad guy" only wants to *read* the message and against the **active** attack, where the bad guy wants to *alter* the message (including the sender's address). The tools against these attacks are **enciphering** and **authenticating**.

Let $\mathbf{P}$, $\mathbf{K}$ and $\mathbf{C}$ be sets (whose elements will be called **plaintexts (cleartexts)**, **keys** and **ciphertexts**, respectively). A **symmetric enciphering algorithm** consists of a family $f = \{f_K \mid K \in \mathbf{K}\}$ of functions from $\mathbf{P}$ into $\mathbf{C}$ such that for any $K \in \mathbf{K}$ there exists a function f'_K from $\mathbf{C}$ into $\mathbf{P}$ such that $f'_K \circ f_K = id_{\mathbf{P}}$.

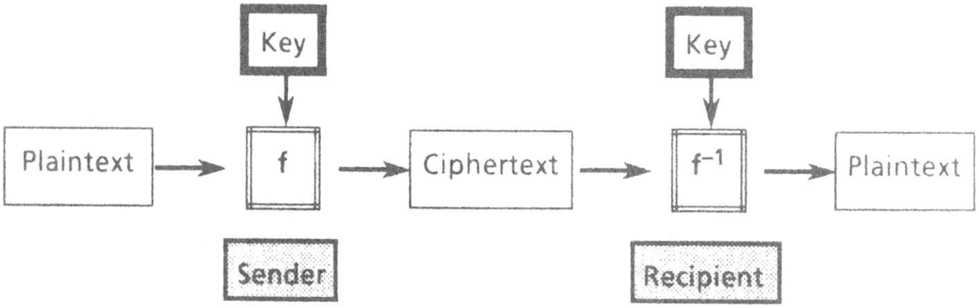

This seems to be difficult, but it is exactly what one expects: The sender **enciphers** a plaintext under a key, and the recipient is able to **decipher** the ciphertext using the same key and will end up in the original plaintext.

In order to estimate the security of an algorithm, one usually accepts Kerckhoffs's principle, that the security must not rely on the fact that the algorithm is kept secret. On the other hand, the actually used key must be kept secret, that is it must be known only to sender and receiver.

An authentication algorithm can be described similarly. Again, let $\mathbf{P}$, $\mathbf{K}$ and $\mathbf{C}$ be sets (whose elements are called **plaintexts, keys** and certain "control data", also called **message authentication codes**). A (symmetric) **authentication algorithm** consists of a family $f = \{f_K \mid K \in \mathbf{K}\}$ of functions from $\mathbf{P}$ into $\mathbf{C}$.

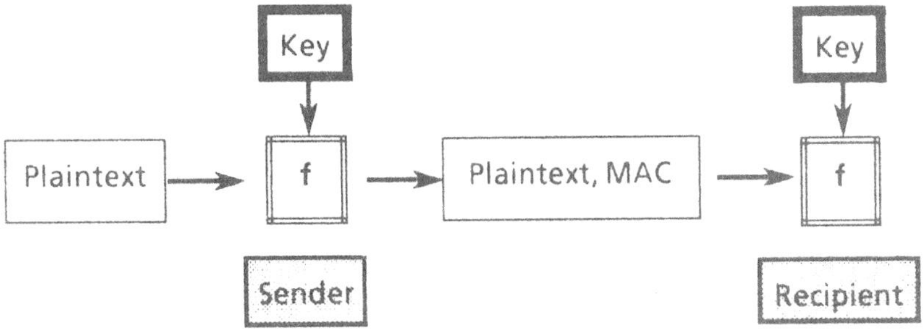

The explanation is as follows. The sender sends along with the actual message the message authentication code; the recipient verifies whether the received message and the received message authentication code fit.

We note an extremely important observation: For both types of algorithms is it necessary that sender and recipient share a common secret key. This fact has many consequences, a few of which are the following.

- Every pair of participants who want to communicate using a (symmetric) crypto algorithm has to have a common secret key. Therefore, one needs a huge amount of secret keys.
- Every secret key has to be sent at least once. (We do not consider a Diffie-Hellman type of key-exchange.)
- Any participant has to store his secret keys in such a way that they cannot be read out. This is a serious problem, since those keys should (ideally) be hardware protected.

All these problems belong to the problems of key management. We shall meet some of these problems in the paper and shall then discuss the possibility of solutions via finite geometry.

The reader who is interested in cryptology is referred to the literature [3], [8], [15], [18], [32], whereas the geometric background can be found in [4], [5], [17], [22], [23] and [31].

1.2 Geometry and Cryptography

1.2.1 Why Geometry?

Why are finite geometric structures applicable? Is this a shere miracle or is it possible to provide reasons for the fact that geometry (geometries) can be applied?

We will give two answers to this question — one is concerned with the word "finite", the other with the word "geometry".

First, we observe that the majority of todays applications deals with *discrete structures*, and not (as it has been till 40 years ago) with continuous structures exclusively. To put it plain, physics is no longer *the* field of applications of mathematics, but rather computer science, social sciences, and so on. These are sciences which deal with individuals, that is they study discrete structures. From a mathematical point of view, analysis is no longer *the* model for applications, but rather combinatorics, or discrete mathematics, in particular (parts of) algebra, number theory, statistics, combinatorics — and geometry.

Second question: Is there any advantage that geometry has? In other words: Are there instances in which geometry is more suitable than (for instance) pure combinatorics or number theory?

In my opinion, there are two points in favour of geometry. Firstly, it turns out that geometries provide examples in the "extreme" cases. This reflects the underlying believe that extreme cases are beautiful. We shall see this phenomenon when we discuss authentication schemes.

Secondly, and more importantly, geometries can be used as models for very complex structures. In particular, hierarchical structures can be modelled by geometries most efficiently. This shall be seen in particular in the fields of threshold schemes and WOM-codes.

1.2.2 Which Parts of Geometry are Applicable?

What is being applied? Only definitions ??? Theorems?? Structures?

As a matter of fact, (nearly) no theorems of discrete mathematics are applied, but (nearly) exclusively *structures*. So, the geometry is not "lost", when the application has been established, but it remains within the application. Needless to say that there is also a strong influence from the applications on geometry. It happened quite a few times (and in future it will happen more often) that a purely "applied" question has a translation into a geometric question, which is by no means trivial.

Let us ask a further question. *What kind of geometries does one use?*

The surprising answer is that only the very *classical structures* are being applied, that is mainly projective spaces, i.e. geometries over (finite) fields. On the other hand, "exotic structures" interesting as they are, have found no application so far. Why? The answer is simple. In order to apply a structure, one must be able to compute very fast in these structures. This is only possible if the underlying arithmetic is very well understood, which is virtually only the case for fields.

We shall now survey selected areas of applications of geometry to cryptography, namely authentication schemes, threshold schemes, communication networks and WOM-codes.

2. Authentication Systems

Suppose that Caesar wants to send a message M ("Ti amo") to Cleopatra. It is important that Cleopatra receives the message without any alteration. On the other hand, a bad guy **X** looks for his chance to alter M in his favour. In order to make the bad guy's life difficult, Caesar **authenticates** the message M.

Therefore, Caesar and Cleopatra have to agree on an authentication function f and a secret key K. The function f has M and K as its input, and the **authenticator** (also called **message authentication code**) f(M,K) as its output.

Now the procedure is as follows. Caesar sends the message M along with the authenticator A = f(M,K). Cleopatra receives a message, say M' and an "authenticator" A'. She computes A* = f(M',K). Only if A* = A' she accepts the received message as it stands.

Here we shall deal only with **unconditionally secure** authentication systems. This means that the security of the system is only based on the number (and distribution) of the keys and does not depend on the security of a crypto-algorithm (those systems one would call **computationally secure**).

In the literature one finds also frequently the expressions **source states, encoding rules, messages** for messages, keys and authenticators, respectively.

What can a bad guy do? He wants to delete M and to insert another message M* ("Ti odio"). Since he does not know the secret key K, he has no method to forge M, he can only try. This stupid method shows that the probability of cheating is at least 1/k, where k denotes the number of keys. But the bad guy's chances of success are not as bad as it may seem. Gilbert, MacWilliams and Sloane [20] have proved the following

2.1 Theorem. *Suppose that any authenticator has just one message. Assume furthermore that all messages and all keys occur with the same probability. Denote by* k *the total number of keys. Then, in any authentication system, the bad guy's chance of success is at least* $1/\sqrt{k}$.

An authentication system in which the bad guy's chance is exactly $1/\sqrt{k}$ is called **perfect**. In other words, in a perfect system the chance of success for a bad guy is as small as one can hope for. Gilbert, MacWilliams and Sloane [20] have constructed perfect authentication systems using projective planes.

2.2 Example. Consider a projective plane **P** and fix a line ℓ in **P**. Define *messages* to be the points on ℓ and *keys* the points off ℓ; the *authenticator* belonging to a message M under the key K is simply the line through M and K. Hence, the authenticators are exactly the lines different from ℓ.

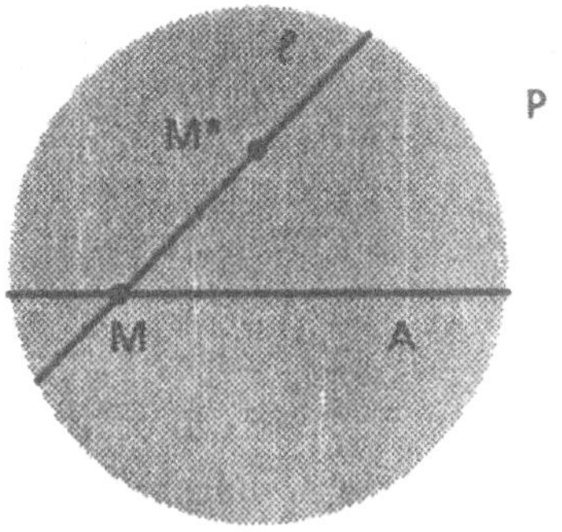

If **P** has order n, then there are exactly $k = n^2$ keys. Consider now the possibilities for the bad guy He sees a message M along with its authenticator A. In

order to insert a different message M^* successfully, he has to find the correct authenticator. This is the line connecting M^* with the key — which is one of the n affine points on the line A. So, the bad guy cannot do better than choosing one of theses points (or, equivalently, one of the n lines $\neq \ell$ through M^*) at random. In other words, his chances of success are precisely $1/n = 1/\sqrt{k}$. Hence, the system is perfect.

These examples, beautiful as they are, have two disadvantages:

- there are very few messages (compared with the number of keys).
- If the bad guy happens to know two messages authenticated with the same key, he knows the key.

In order to deal with these problems, one can prove the following theorems (see [12] and [36]).

2.3 Theorem. *Let* **A** *be a perfect authentication system with* $k = k_1^2$ *keys. Denote the number of messages by* m *and the number of authenticators by* a. *Then* $m \leqq k_1 + 1$ *with equality if and only if* **A** *is the example* **2.2**.

We can construct authentication systems with 'many' messages. In view of the above theorem, they cannot be perfect in a strong sense, but **essentially perfect**. By this we mean that the bad guy's chance of success is only $O(1/\sqrt{k})$. The known constructions are based on geometric structures, in particular finite projectivce spaces, see [7], [12] and [20]. The reader may also want to look at the papers [37] and [38], where authentication schemes are constructed from combinatorial structures.

In order to overcome the second difficulty, we construct the following examples:

2.4 Examples [12]. *Let* $\mathbf{P} = PG(s+1,q)$ *be the projective space of dimension* $s+1$ *of order* q. *Fix a hyperplane* H_0 *in* **P** *and choose a set* $\mathcal{M}$ *of* $(s-1)$*-dimensional subspaces of* H_0. *Define the authentication system* **A** *as follows:*

Messages of **A** are the elements of $\mathcal{M}$;

the *keys* of **A** are the points of **P** outside H_0;

the *authenticator* through message M and key P is the hyperplane spanned by M and P.

(Note that in case $s = 1$ we get example **2.2** back.)

Clearly, the bad guy has an *a-priory chance* of forging the system of $1/q$. (If he wants to substitute the message M^*, he chooses as an authenticator one of the q hyperplanes $\neq \mathbf{H_0}$ through M^* at random.)

We can convince ourselves that the bad guy's chance of success does not increase considerably, even if he knows s authenticators computed with the same key. This observation is based on the fact that s authenticators determine only a subspace of dimension $s-1$, through which there are still q hyperplanes (possible keys) at disposal.

More precisely: If M consists of hyperplanes in general position (that is no $s+1$ of them have a point in common), then the bad guy's chance of success after having observed s authenticators is still only $1/q$. Such authentication systems are called **s-fold secure** [12]. In [12] and [13] there are constructed many **essentially s-fold secure** authentication systems, which means that if the bad guy's chance of success is at $1/q$ at the beginning, it is still only $O(1/q)$ after he has observed s authenticated messages.

3. Threshold Schemes

The usual way to control the execution of an operation which requires a quorum or threshold of t members of a group of n "authorised users" is to hold a meeting at which the members are physically present. Sometimes, however, the authorised users cannot identify each other by inspection; a meeting might not be feasible or the users do not know each other personally (for example, some of them might be computer programs). In order to achieve a quorum in such a situation, so-called threshold schemes have been introduced (see, for instance, [2, 14, 24, 33, 34, 35]). A good name for threshold schemes (and "generalized" threshold schemes we shall discuss below) is also **secret sharing schemes**.

A **t-threshold scheme** consists of $n \geq t$ pieces of information (sometimes called *shadows*) such that the following properties are satisfied:

(i) a secret datum X can be retrieved from any t shadows,

(ii) X cannot be determined from any $t-1$ or fewer of the shadows.

There are many examples of threshold schemes, most of them constructed using finite geometry. Geometric threshold schemes have additional advantages. For instance, it is possible to introduce a hierarchy in the set of shadows (**multilevel** schemes), or to divide the shadows into independent groups (**compartmented** schemes). We shall also exhibit systems which reflect the fact that in many situations some users have more rights than others.

In terms of geometric language, a t-**threshold scheme** can be described as follows. In a *geometry* consisting of *points* and *blocks* one chooses a block B and n points on B in such a way that

(i) any t of the n points determine B uniquely,

(ii) there are 'many' blocks through any $t-1$ or fewer of the n points.

Thus the shadows in a geometric threshold scheme correspond to points of a (randomly chosen) block which can be retrieved from any t but not from fewer than t of its points.

The following realization has the property that the system has to store only extremely few secret information. We select a set S of points which intersects the block B in a unique point X (or a non-empty set X of points) and store S and X. If u points enter the system, it tries to construct a block through these points. Only if there is a unique block C, say, which contains the u points, the system computes $C \cap S$. The threshold is achieved if and only if $C \cap S = X$. This way we don't need to store any information about the block B, it suffices to make X inaccessible.

What happens if an unauthorised user takes part in the process? Even if there is a unique block C through the points entered we can virtually be certain that this block is different from B and that $C \cap S$ is not the point (or set of points) X. So the existence of only one unauthorised user U makes sure that no group to which U belongs can perform the critical operation.

The precise meaning of 'many' in above description depends of course on the particular scheme. We shall see that we can achieve any level of security by choosing the geometry accordingly.

Threshold schemes have been introduced by Shamir [33] and Blakley [2]. In his paper [33] Shamir describes one class of examples which is still the prototype of all threshold schemes.

If the order of the underlying projective space is q, then the probability of cheating is in almost all of our examples $1/(q+1)$.

3.1 Example [33]. *Choose a polynomial* $f = a_0 + a_1x + a_2x^2 + \ldots + a_{t-1}x^{t-1}$ *of degree* $t-1$ *at random. Let* B *be the set of all points* (x,y) *in the plane which satisfy* $f(x) = y$. *The actual secret* X *is* a_0, *in other words, the set* S *is the* y*-axis.*

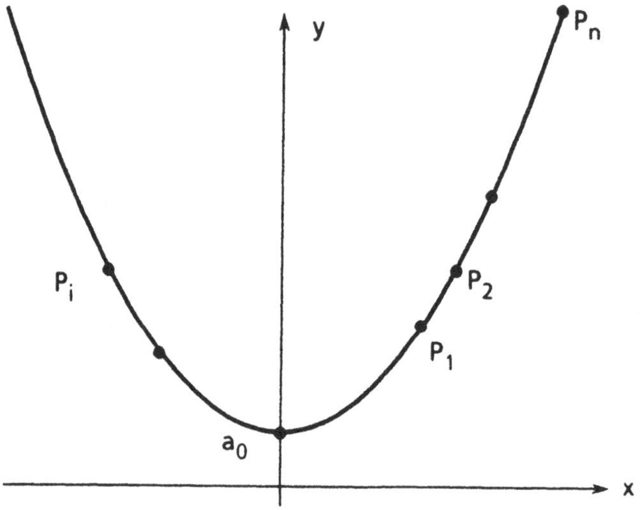

Example. t = 2

The example is based on the fact that f is uniquely determined by any t of its points. (Actually, the polynomial f can easily be retrieved from any t of its points using, for instance, interpolation.) On the other hand, given $t-1$ or fewer points outside the y-axis, any hypothetical secret $X' = (a_0',0)$ lies together with the $t-1$ points on a polynomial of degree $\leqq t-1$.

We shall here discuss another example, which is, in a way, more flexible for different applications.

3.2 Example. In a projective geometry $\mathbf{P} = PG(d,q)$ we choose a (t−1)-dimensional subspace B and a line ℓ intersecting B at a point X. Furthermore, we choose as shadows n points $P_1, P_2, \ldots, P_n$ of B such that the set $\{X, P_1, P_2, \ldots, P_n\}$ is in general position, that is, any t of these points generate B.

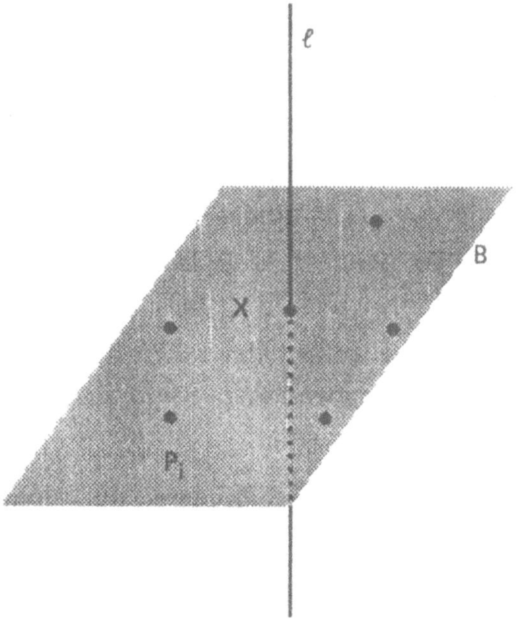

Clearly, this gives a t-threshold scheme. In particular we see that *there is a t-threshold scheme for any positive integer* t with an unlimited number of possible shadows.

Using this construction method, one can easily obtain other schemes.

3.3 Multilevel schemes

In reality, some people are more equal than others. In other words, there is a hierarchy. One knows this from the military world (a general has more power and responsiblity than a colonel, who has more influence than an ordinary soldier). The same principle holds also in the commercial world. In the last years, however, I have encountered this phenomenon very rarely in the academic world. (I'm only observing it, I don't complain.)

Such a hierarchy can be modelled easily in any geometry. Every user gets as his shadow not necessarily a point, but a subspace whose dimension corresponds to his rank in the hierarchy.

3.4 Compartmented schemes

The above mentioned scheme has the property that a user of high rank can be substituted by (many) users of low rank. This is sometimes an advantage, sometimes not. In a compartmented scheme there are several groups G_i ($i = 1,\dots,m$), in every group one has a threshold t_i and the global secret can be constructed only if in a certain number t_∞ of groups the respective quorum is achieved.

This can, for instance, be modelled as follows. First, for every group G_i one defines a linear subspace U_i of dimension t_i-1 of a projective space P. These subspaces U_i are chosen in such a way that

- they are in general position, that is they generate a space of dimension $t_1 + \dots + t_m - 1$,

- they intersect a subspace U_∞ of dimension $t_\infty-1$ in points $X_1, \dots, X_m$, respectively,

- the points $X_1,\dots,X_m$ are in general position in U_∞.

Now the procedure is as follows. First the system tries to construct the secret X_i in any group G_i. Then, with the so-obtained local secrets X_i of U_∞, the global secret X which is the intersection of U_∞ with a line S is constructed in the usual way.

For further references see [11], [34] and [35].

4. Communication networks

4.1 How to communicate efficiently.

Consider a distributed system consisting of a set of N autonomous computers interconnected by a communication network. We suppose that any single transmitted message of one computer to another adds to the costs of the global communication. We shall simply speak of a network with N **knots**.

The problem we want to consider is known as "decentralized consensus problem". There is some information gathered from all knots, this information is synthesized and the outcome must be made known to all knots. An example is that the knots are machines and status information of the machines must be known to every knot. We want to develop an algorithm which guarantees that after some time every knot of the net knows the critical value. This means that "in a way" any value must reach every participant. Such a procedure was called [1] "broadcasting without broadcast".

Clearly, such algorithms exist. Here is one: Every knot sends his data to every other knot. Then every knot can evaluate the received data and knows the outcome. Since there are N knots, one needs a total of $N(N-1)/2$ transmit operations. Hence the complexity of this algorithm is $O(N^2)$.

The question is: What are the most efficient algorithms ? Given N, we denote by t_N the number of transmits of the best algorithm which satisfies the rules. Hence the above algorithm implies $t_N \leqq c \cdot N^2$.

Using geometric methods (in particular projective planes) it was proved in [25] that $t_N \leqq c \cdot N\sqrt{N}$.

In a remarkable paper [1], Alon, Barak and Manber proved (in a rather tricky way) that $t_N \leqq c \cdot N \cdot \log_2 N$.

The aim of this note is to prove the same result in an very easy way using finite geometry. Our main result reads as follows

Theorem. *Fix a prime-power* q. *Let* d *be the smallest positive integer with* $N \leqq q^d$. *Then* $t_N \leqq c \cdot N \cdot d$, *where* c *is a constant depending on* q. Hence $t_N \leqq c \cdot N \cdot \log_q N$.

As the algorithms in [1] and [25], our algorithm is "completely symmetric", that is every knot has the same duties

The idea of the *proof* is as follows.

Let **A** = AG(d,q) be the affine space of dimension d of order q. We denote the hyperplane of infinity by H_∞. Let $P_1, \ldots P_d$ be d points of H_∞ in general position.

Then these points generate H_∞. Let $\Pi_1, \ldots, \Pi_d$ be the parallel classes (of lines) in **A** such that the lines of Π_i intersect in P_i ($i = 1, \ldots, d$).

We identify the knots of the network with a subset of the points of **A**.

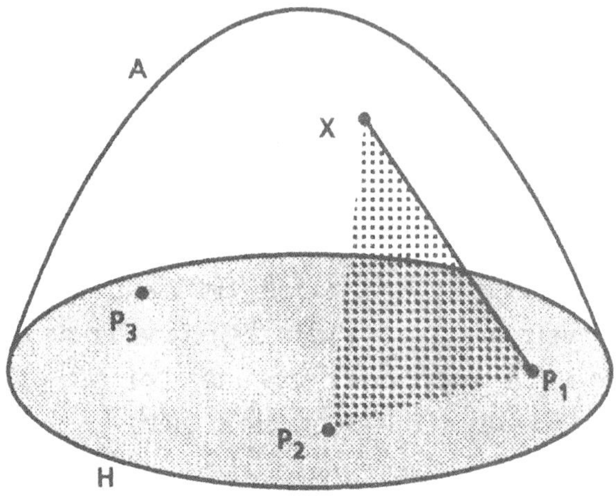

Example. The case d = 3

The algorithm runs jn d steps. In step i every point P sends its data to all the affine points of the line PP_i. One shows easily that after i steps the critical data are know to all points of subspace of dimension i. In particular, after d steps, the message will be distributed among all knots.

For details see [9].

4.2 Switching Networks

Imagine a telefone network with N participants. These participants are connected via switches in the following way.

Every participant P is connected to a certain number, say k, of switches. Of course, this set of switches must have the property that every participant $\neq$ P is connected to at least one switch, P is connected to. In the following figure one can see an example. The "points" on the outer circle are the N = 7 participants, the "points" on the inner circle are the switches.

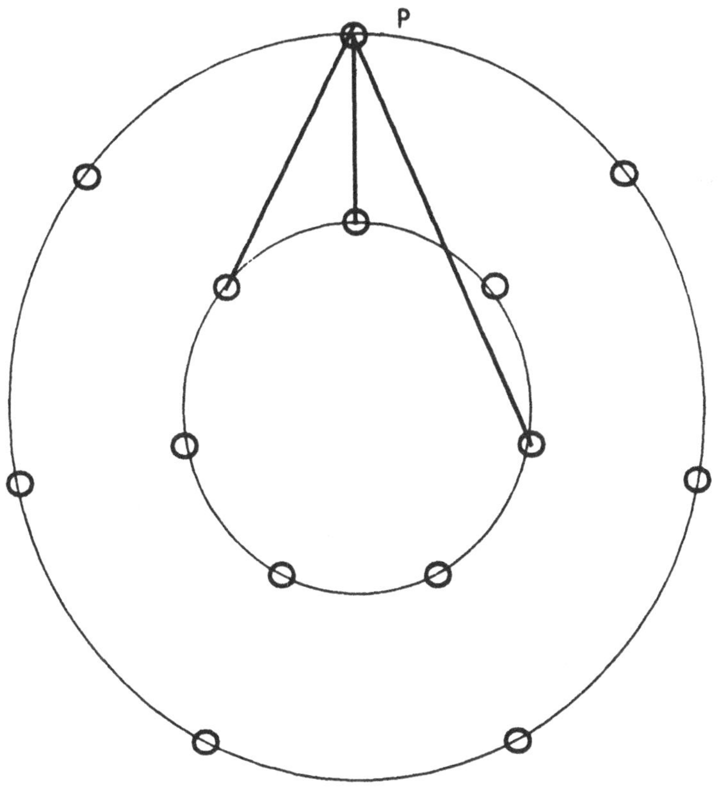

One would like to design such a system with the following properties:

- The number of switches is minimal.
- All switches are alike. (This is necessary since the switches are being fabricated in mass production.)

A very effective system was proposed by Hagelbarger [21]. Let N be of the form $N = q^2 + q + 1$, where q is a power of a prime. Identify the participants with the points of the Desarguesian projective plane $PG(2,q)$.

It is well known that $PG(2,q)$ can be constructed using a (cyclic) difference set. This means the following. We identify the points of $PG(2,q)$ with the elements of $\mathbf{Z}_N$. There exists a difference set $D \subseteq \mathbf{Z}_N$, which can be defined by the property that the lines of $PG(2,q)$ are precisely the translates $D + x$ $(x \in \mathbf{Z}_N)$ of D.

Now the rule is simple. The number of switches equals the number N of participants. More precisely, we identify the switches with the lines of PG(2,q). A participant P_i is directly connected to the switch s_j if the point P_i is on the line s_j of PG(2,q).

As an example, take N = 7, that is q = 2. A difference set of PG(2,2) is {1,2,4}. The corresponding system looks as follows.

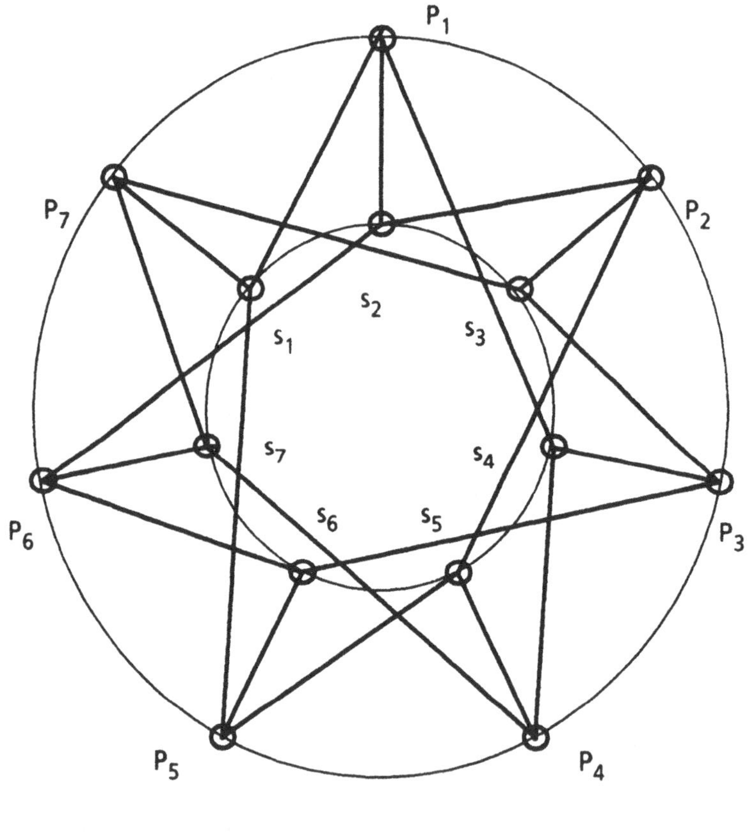

4.3 Reducing the storage amount of communication keys

If any two participants in a system of N participants want to be able to communicate with each other using a symmetric cryptosystem, one needs a total of N(N−1)/2 secret keys. We call this procedure here the **naive** approach. In particular, any participant has to store N−1 keys in a secret way. Since storing secret data is a serious problem (usually one needs special physically protected devices), one tries to reduce the amount of secret data.

The idea is that any participant P stores a ("small") set of "subkeys". If P, and Q want to communicate, they "combine" their subkeys and compute the communication key. Geometric designs may help to construct such systems. The basic idea for this is as follows.

Consider a geometric design **D** consisting of "points" and "blocks". We may assume that each block is a subset of the points set of **D**.

The participants are identified with the points of **D**. The set of subkeys of the system will be indexed by the set of blocks of **D**. (Each subkey belongs to exactly one block.) This is done in such a way that the set of subkeys belonging to participant P is indexed by the set of blocks through the point P. Note that the structure of the geometry is public, whereas the subkeys are secret. This means that a subkey is known only to a point (participant) which is incident with a block which is the index of the subkey under consideration.

If participants P and Q want to communicate, they take the blocks through P and Q and look for the corresponding subkeys (which are known to both of them). Then they apply an algorithm to the set of their common subkeys. For instance, this operation may consist of taking the sum, or the union of the common subkeys. We call this structure the **key distribution pattern with respect to D**, KDP(**D**), for short.

If for example, D is a 2-(v,k,λ) block design with r blocks per point, then any participant has to store only r subkeys compared to v−1 keys in the naive case. So in order to reduce the amount of storage, one should use block designs with "small" r

Yet, there is another problem. In the naive setting, the N(N−1)/2 keys have been chosen independently; so no coalition of participants can compute the key of another participant, even if they pool their keys.

In the model of a KDP(**D**), the situation is quite different. Consider, for instance, the case of a 2-(v,k,1) design **D**. In this case any two participants on the same block have equal communication keys. This is a completely unacceptable situation – except in the case k = 2, where the KDP(**D**) reduces to the naive model.

In order to describe the security of such a key distribution pattern, Mitchell and Piper [20] have introduced the following definition. Let w be a positive integer. A KDP(D) is called w-collusion resistant, if the following condition is satisfied:

For any w-element subset W of the point set and for any two points $P, Q \notin W$ it holds

$$(P) \cap (Q) \not\subseteq \bigcup_{X \in W} (X),$$

where (Y) denotes the set of all blocks through the point Y.

In other words, a w-collusion resistant KDP(**D**) has the property that in the pool of the keys belonging to the "bad guys" in W, the communication key of P and Q is not contained.

The particular case of 3-designs is studied in the following

Lemma [20]. *Let* **D** *be a* 3-(v,k,λ) *design and denote by* λ_2 *the number of blocks through two points. (Then* $\lambda_2 = (v-2)\lambda/(k-2)$.*) Then the* KDP(**D**) *is* w-*collusion resistant with* $w \geqq \lambda_2/\lambda$.

The *proof* is easy: In order to obtain the communication key between P and Q the bad guys have to know the λ_2 blocks between P and Q. Each bad guy X knows the λ blocks through P, Q and X. So in order to obtain λ_2 blocks, at least λ_2/λ bad guys have to make a coalition.

This Lemma shows that one should look among the 3-designs for designs with "big" λ_2/λ.

Let us finally discuss the situation for 3-(v,k,λ) designs. They satisfy

$$\lambda_2 = (v-2)\lambda/(k-2) \text{ and } r = (v-1)(v-2)\lambda/(k-1)(k-2),$$

hence

$$\lambda_2/\lambda = (v-2)/(k-2).$$

Suppose we have a network with v participants. In order to make λ_2/λ big, one has to make k small. On the other hand, in order to make r small, one has to choose λ small and k big.

A good example which was considered in the literature are the **Möbius planes**, in other words, the $3\text{-}(q^2+1,q+1,1)$ designs. These structures exist for any prime-power q. They satisfy $\lambda_3 = 1$, $\lambda_2 = q+1$ and $r = q^2+q$. It follows that the corresponding KDP is at least (q + 1)-collusion resistant.

For more details see [27], [28], [29].

5. WOM-Codes

Many modern storage media are made of write-once-memories (WOMs). In these memories each cell has an initial value (for instance 1), and any cell can be written only once. That is, its value can be put 0, but if a value is 0 it will stay 0 forever.

The reason for the usage of these storage media is clear. Storing data in a WOM is a physical protection against alteration of the data. If such a memory is delivered to the end-user, its content shall be unchangable. The today most popular form of write-once-memories are the compact discs.

For us, a good model of a WOM is a piece of paper, in which at certain positions, a hole can be made; of course, a hole cannot be unmade. So, the good-old punched cards are write-once-memories.

How does one use such a memory? In order to introduce the (mathematical) problems, we consider a simplified example. Imagine that we have 25 cells. In order to be able to speak easily, we associate to each cell a letter of the alphabet. (See the following figure.)

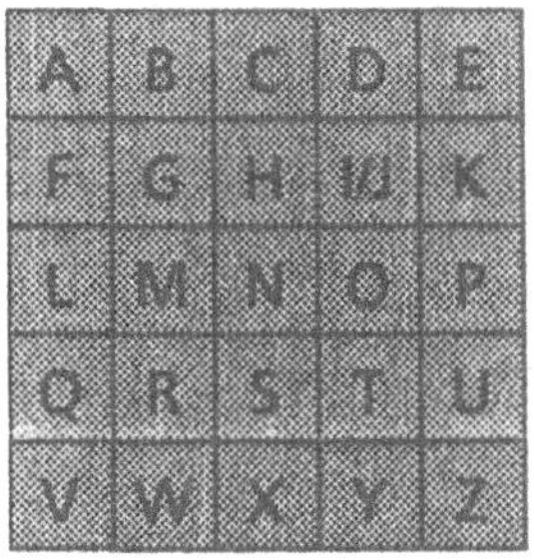

Each cell has originally value 1 (is black) and can become white (get value 0). In this way, one can write a "message" onto the medium. For instance, we can write a letter (see the following figure).

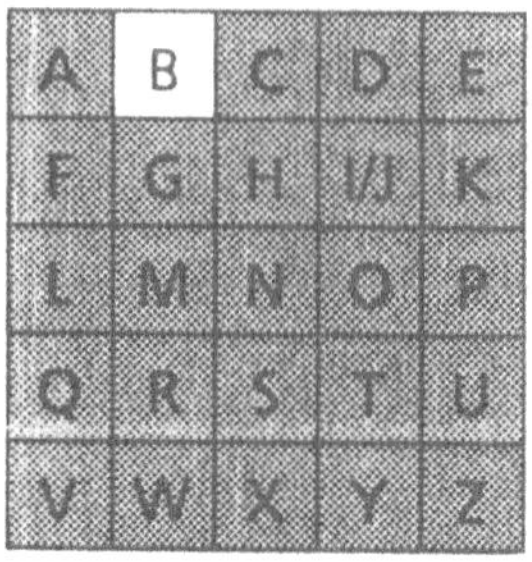

More precisely, we can write *one* letter. Why? If the memory has been read and we would like to use it again, we may write another cell,

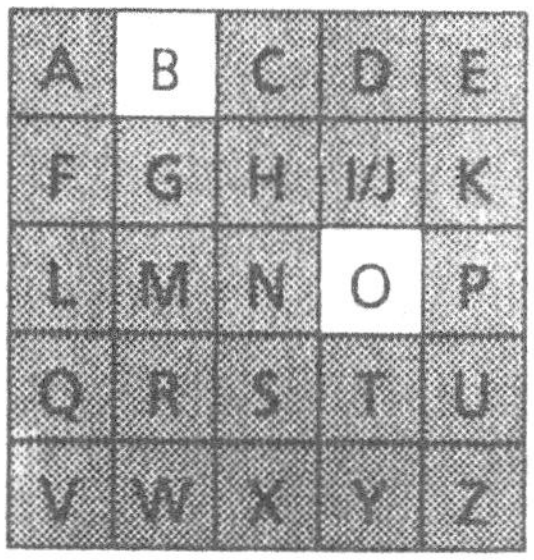

– but now it is totally unclear to the receiver which cell we have "meant"; in our example, he has the choice between B and O.

The problem becomes even more intricate if we consider a third turn. Suppose that, by chance, the third letter equals the first. Since the corresponding cell has already been written, by definition of a WOM, then there is no possibility at all to indicate which letter we do mean now. (One case is easy (and trivial): If a letter is to be repeated, we let the memory as it is.)

So, the problem is the following: Is there any chance to write an *arbitrary* sequence of letters with prescribed length? Of course, one could reserve several cells for the most popular letters, and so, statistically, one could write rather long messages. But what about the worst case? To put the above question a little bit different:

1. *Can one organize the set of cells in such a way that any sequence of letters of a certain length can be written?*

But, as we have seen above, there is a much more serious problem. A WOM-reader wants to read the latest message (in our example, the latest written letter). Moreover, the reader must be able to do this without knowing the history of the medium (that is, the earlier written letters). In other words, such a reader is in the situation of the above example. He sees certain sets of black and white cells and has to know which message the medium means. For this, the reader must recognize a pattern.

So, we have a second condition for writing onto a write-once-memory.

2. *Any admissible pattern of black and white cells must define uniquely one cell.*

The main idea to solve these problems is the observation that no cell is an island. If we provide structure on the set of cells, a certain cell may be marked not only by coloring this very cell black, but also by coloring the neighborhood of this cell. In particular, a cell may be distinguished without coloring it black. (We may hit a cell without killing it.)

In order to do this, the set of cells has to have a certain structure. And here, as a surprise, geometry, in particular finite geometry, comes in very naturally!

A particular beautiful example is as follows (cf. [26], [10]). Let $\mathbf{S}$ be a Steiner triple system. We start as in 3.1: $\mathbf{B}_0$ consists of the singletons and $\mathbf{B}_1$ of the punctured lines, which are in this case the unordered pairs of points.

1. Generation:

2. Generation:

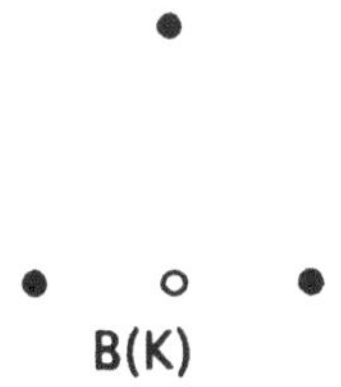

As $\mathbf{B}_2$ we now take the sets K of 4 points consisting of the sets of points on a line and one point P(K) outside this line.

3. Generation:

It is easy to see that in all situations, a pattern of $\mathbf{B}_1$ can be extended to a pattern of $\mathbf{B}_2$.

For the fourth generation $\mathbf{B}_3$ we take two types of sets K. Either tne union of two intersecting lines (in which the point of intersection is P(K)) or the set of all points with one exception – and, of course, the exceptional point is distinguished!

4. Generation:

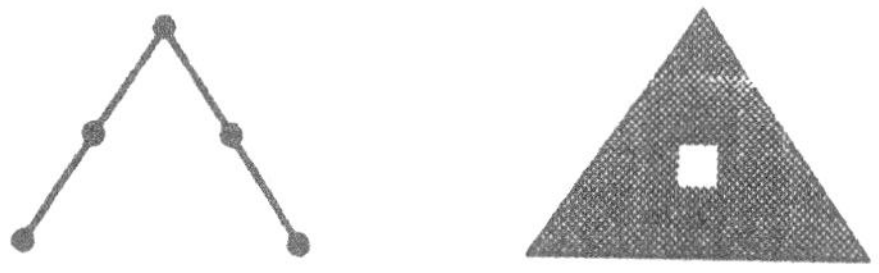

This procedure applies to every Steiner triple system, even to the two smallest (and most famous) ones, the projective plane of order 2 (**Fano plane**)

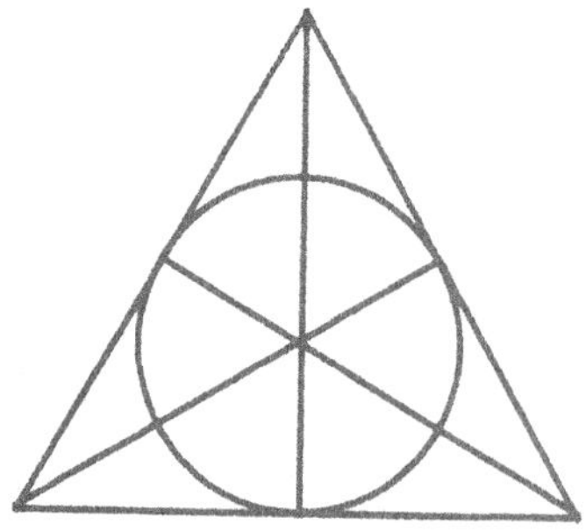

and the affine plane of order 3:

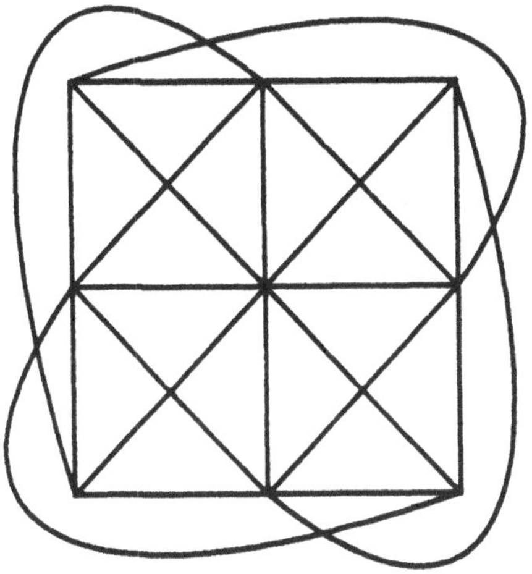

In order to write a message consisting of several letters which should be read simultaneously, we use a series of WOMs. Let us consider an example, where we use the WOM-code obtained above with an affine plane of order 3, where the assignment of the letters to the 9 points is as follows:

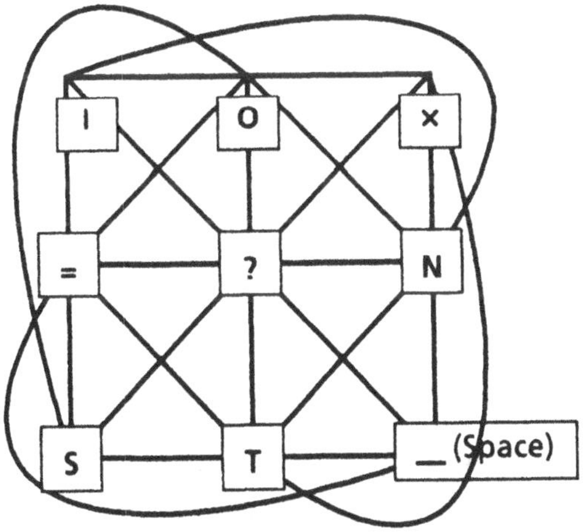

The reader may decode the below messages. (Every reader of this paper should be able to understand the mathematics involved!)

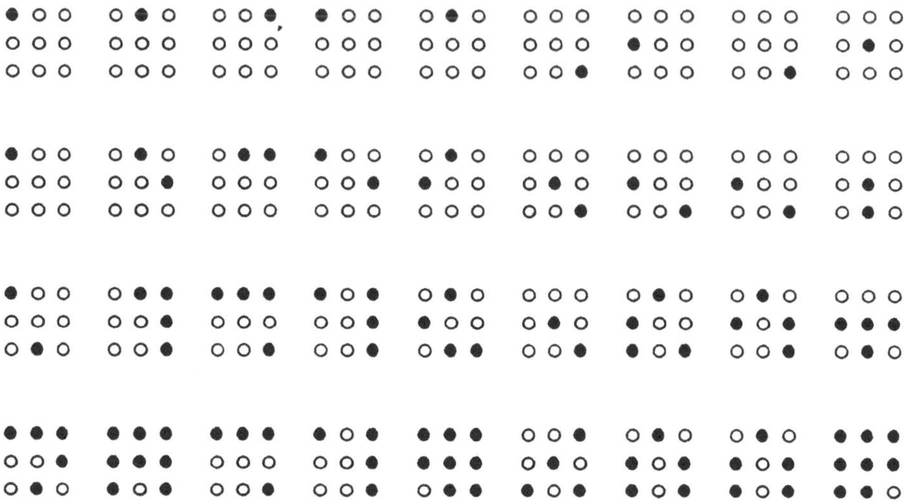

WOM codes have been considered in [16], [19], [26], [30] and [39]. The connections with geometry are studied in [10].

Acknowledgement

I would like to thank many friends for many discussions on many topics concerning this paper.

References

1. Alon, N., Barak, A., Manber, U.: *On Disseminating Information Reliably Without Broadcasting*, IEEE, The 7th international conference on distributed computing systems, Berlin, Sepember 21 - 25, 1987.
2. Blakley, G.R.: *Safeguarding cryptographic keys*, Proc. NCC **48**, AFIPS Press, Montvale, N.J., 317-319 (1979).
3. Beker, H. and Piper, F.C.: *Cipher Systems*, Northwood Books, London 1982.
4. Beth, T., Jungnickel, D., Lenz, H., *Design Theory*, B.I.-Wissenschaftsverlag, Mannheim - Wien - Zürich, 1985 and Cambridge University Press, Cambridge, 1987.

5. Beutelspacher, A., *Einführung in die endliche Geometrie I. Blockpläne*, B.I.-Wissenschaftsverlag, Mannheim - Wien - Zürich, 1982.

6. Beutelspacher, A.: *Enciphered Geometry*, Some Applications of Geometry to Cryptography. Annals of Discrete Math. **37** (1988), 59-68.

7. Beutelspacher, A.: *Perfect and essentially perfect authentication systems.* Extended Abstract, Advances in Cryptology: Proceedings of EUROCRYPT 87 (Lecture Notes in Computer Science **304**), D. Chaum and W.L. Price, Eds., Springer-Verlag 1988, 167-170.

8. Beutelspacher, A.: *Kryptologie*, Vieweg-Verlag, Braunschweig und Wiesbaden 1987.

9. Beutelspacher, A.: *How to communicate efficiently,* to appear in J. Combinat. Th. A.

10. Beutelspacher, A.: *How to use a write-once memory as often as possible*, manuscript.

11. Beutelspacher, A., Vedder, K.: *Geometric structures as threshold schemes.* The Institute of Mathematics and its Applications, Conference Series **20**, Cryptography and Coding (ed. by H.J. Beker and F.C. Piper), 1989, 255-268.

12. Beutelspacher, A., Rosenbaum, U.: *Geometric authentication systems*, to appear in Ratio Mathematica.

13. Beutelspacher, A., Tallini, G., Zanella, C.: *Examples of essentially* s-*fold secure geometric authentication systems with large* s, to appear in Rend. Mat. Roma.

14. Chaum, D.: *Computer systems established, maintained, and trusted by mutually suspicious groups*, Memorandum No. UCB/ERL M79/10, University of California, Berkeley, CA, February 22, 1979.

15. Davies, D.W., Price, W.L.: *Security for Computer Networks*, John Wiley & Sons, 1984.

16. Cohen, G.D., Godlewski, P., Merkx, F.: *Linear Binary Code for Write-Once Memories*, IEEE Trans. Inform. Theory **32** (1986), 697-6700.

17. Dembowski,P.: *Finite Geometries*, Springer-Verlag, 1968.

18. Denning, D.E.R: *Cryptography and Data Security*, Addison-Wesley, Reading, Mass., 1983.

19. Fiat, A., Shamir, A.: *Generalized "Write-Once" Memories*, IEEE Trans. Inf. Theory IT-30, 1984, 470-480.

20. Gilbert, E.N., MacWilliams, F.J., Sloane, N.J.A.: *Codes which detect deception*, Bell. Syst. Tech. J. **53** (1974), 405-424.

21. Hagelbarger, D.W.: *The application of balanced symmetric incomplete block designs to switching networks*, Proceedings of the international conference on communications, June 14, 15, 16, 1971, Montreal.

22. Hirschfeld, J.W.P.: Geometries over finite fields, Clarendon Press, Oxford 1979.

23. Hughes, D.R., Piper, F.C.: *Design Theory*. Cambridge University Press, Cambridge 1985.

24. Kothari, S.C.: *Generalized Linear Threshold Scheme*, Advances in Cryptology (Proceedings of CRYPTO 84), Lecture Notes in Computer Science **196**, Springer 1985, 231-241.

25. Lakshman, T.V., Agrawala, A.K.: *Efficient decentralized Consensus Protocols*, IEEE Transactions on Software Engineering, **12** (198), 600-607

26. Merkx, F.: *Womcodes constructed with projective geometries*, Traitement du signal **1** (1984), 227-231.

27. Mitchell, C., Piper, F.: *The cost of reducing key storage requirements in secure networks*, Computers and security.

28. Mitchell, C., Piper, F.: *Key storage in secure networks.*

29. Mitchell, C.: *Combinatorial techniques for key storage reduction in secure networks.*

30. Rivest, R.L., Shamir, A.: *How to reuse a "Write-once" Memory*, Information and Control **55** (1982), 1-19.

31. Segre, B.: *Lectures on modern geometry*, Cremonese, Roma 1961.

32. Sgarro, A.: *Crittografia*, Franco Muzzio, Padova.

33. Shamir, A.: *How to share a secret*, Comm. ACM Vol. **22** (1), 612-613 (1979).

34. Simmons, G.J.: *How to (really) share a secret*, to appear in Advances of Cryptology - CRYPTO 88.

35. Simmons, G.J.: *Robust shared secret schemes.* To appear in Congressus Numerantium, vol. 68-69.

36. de Soete, M., Vedder, K. Walker, M.: *Authentication schemes constructed from generalized polygons*, to appear in Proceedings of EUROCRYPT '89.

37. Stinson, D.R.: *Some constructions and bounds for authentication codes*, J. Cryptology **1** (1988), 37-51.

38. Stinson, D.R.: *A Construction for Authentication/Secrecy Codes from Certain Combinatorial Designs*, J. Cryptology **1** (1988), 119-127

39. Wolf, J.K., Wyner, A.D., Ziv, J., Körner, J.: *Coding for a Write-Once Memory*, AT&T Bell Laboratories Technical Journal **63** (1984), 1089-1111.

AUTHENTICATION / SECRECY CODES

M. De Soete
MBLE-I.S.G., Brussel, Belgium

ABSTRACT

We deal with codes having unconditional security, which means that the security is independent of the computing power. Analogously to the theory of unconditional secrecy due to Shannon [17] Simmons developed a theory of unconditional authentication [19]. In this paper we give some bounds and constructions for authentication/secrecy codes with splitting, based on finite geometry and combinatorics.

1. INTRODUCTION

Consider a transmitter who wants to communicate a source to a remote receiver by sending messages through an imperfect communication channel. Then there are two fundamentally different ways in which the receiver can be deceived. The channel may be noisy so that the symbols in the transmitted message can be received in error, or the channel may be under control of an opponent who can either deliberately modify legitimate messages or else introduce fraudulent ones. Simmons [19] showed that both problems could be modeled in complete generality by replacing the classical noisy communications channel of coding theory with a game - theoretic noiseless channel in which an intelligent opponent, who knows the system and can observe the channel, plays so as to optimize his chances of deceiving the receiver. To provide some degree of immunity to deception (of the receiver), the transmitter also introduces redundancy in this case, but does so in such a way that, for any message the transmitter may send, the altered messages that the opponent would introduce using his optimal strategy, are spread randomly. Authentication is concerned with devising and analyzing schemes (codes) to achieve this "spreading".

In the model some simplifying assumptions are made. We suppose that the transmitter and receiver trust each other completely and that neither acts to deceive the other. We also assume that only the receiver need be convinced of the authenticity of a message, so there is no third party (arbiter) involved here. In addition, we also agree that all successful deceptions of the receiver are of equal value to the opponent. We have to distinguish the authentication schemes in which the opponent knows the state of source (message authentication without secrecy) from the message authentication in situations in which the opponent is ignorant of the information being communicated to the receiver by the transmitter.

2. A MATHEMATICAL AUTHENTICATION MODEL

In this model (see [19], [20], [21]) there are three participants: a *transmitter*, a *receiver* and an *opponent*. The transmitter wants to communicate some information to the receiver. The opponent wanting to deceive the receiver, can either impersonate the receiver, making him accept a fraudulent message as authentic, or, modify a message which has been sent by the transmitter.
Let S denote the set of k source states, M the set of v messages and E the set of b encoding rules. A *source state* $s \in S$ is the information that the transmitter wishes communicate to the receiver. The transmitter and receiver will have secretly chosen an *encoding rule* $e \in E$ beforehand. An encoding rule will be used to determine the message $e(s)$ to be sent to communicate any source state s. In a model with *splitting*, several messages can be used to determine a particular source state. However, in order for a receiver to be able to uniquely determine the source state from the message sent, there can be at most one source state which is encoded by any given message $m \in M$, for a given encoding rule $e \in E$ (this means: $e(s) \neq e(s')$ if $s \neq s'$).

An opponent will play *impersonation* or *substitution*. When the opponent plays impersonation, he sends a message to the receiver, attempting to have the receiver accept the message as authentic. When the opponent plays substitution, he waits until a message m has been sent, and then replaces m with another message m', so that the receiver is misled as to the state of source. More generally, an opponent can observe i (≥ 0) distinct messages being sent over the channel knowing that the same encoding rule is used to transmit them, but ignoring this rule. If we consider the code as a secrecy system, then we make the assumption that the opponent can only observe the messages being sent. Our goal is that the opponent be unable to determine any information regarding the i source states from the i messages he has observed.

The following scenario for authentication is investigated. After the observation of i messages $M' \subset M$, the opponent sends a message m' to the receiver, $m' \notin M'$, hoping to have it accepted as authentic. This is called a *spoofing attack of order i* [14], with the special cases $i = 0$ and $i = 1$ corresponding respectively to the impersonation and substitution game. The last games have been studied extensively by several authors (see [4], [10], [18], [19].

For any i, there will be a probability on the set of i source states which occur. We ignore the order in which the i source states occur, and assume that no source state occurs more than once. Also, we assume that any set of i source states has a non-zero probability of occurring. Given a set of i source states, we define $p(S)$ to be the probability that the source states in S occur.

Given the probability distributions on the source states described above, the receiver and transmitter will choose a probability distribution for E, called an *encoding strategy*. If splitting occurs, then they will also determine a *splitting strategy* to determine $m \in M$, given $s \in S$ and $e \in E$ (this corresponds to non-deterministic encoding). The transmitter/receiver will determine these strategies to minimize the chance that an opponent can deceive them.

Once the transmitter/receiver have chosen encoding and splitting strategies, we can define for each $i \geq 0$ a probability denoted P_i, which is the probability that the opponent can deceive the transmitter/receiver with a spoofing attack of order i.

In this paper, we consider only codes without splitting. We shall use the following notation. Given an encoding rule e, we define $M(e) = \{e(s) \mid s \in S\}$, i.e. the set of messages permitted by encoding rule e. For a set M' of distinct messages, define also $E(M') = \{e \in E \mid M' \subseteq M(e)\}$, i.e. the set of encoding rules under which all the messages in M' are permitted. Finally we denote by $AC(k, v, b)$ an authentication system with k source states, v messages and b encoding rules.

EXAMPLE [10] Consider a finite projective plane $PG(2,q)$. The source states are defined as the points of a fixed line L. Hence $k = q+1$. The messages are all the lines different from L ($v = q^2 + q$) and the encoding rules are the points not incident with L giving $|E| = q^2$. Giving a source state s it is encoded by the rule e in the message which is the unique line of $PG(2,q)$ joining both. It is easy to verify that this defines a authentication scheme with $P_0 = P_1 = 1/q$.

3. SECRECY

Considering the secrecy properties of a code, we desire that no information be conveyed by the observation of the messages. A code has *perfect L-fold secrecy* (Stinson [21]) if, for every set M_1 of at most L messages observed in the channel, and for every set S_1 of at most $|M_1|$ source states, we have $p(S_1/M_1) = p(S_1)$. This means that observing a set of at most L messages in the channel does not help the opponent to determine the L source states.

On the other hand, a code is said to be *Cartesian* ([4] if any message uniquely determines the source state, independent of the particular encoding rule being used. In terms of entropy, this is expressed by $H(S/M) = 0$. Hence in a Cartesian authentication code there is no secrecy (it has 0-fold secrecy). Note that the example given in the preceding section defines a Cartesian code.

4. BOUNDS ON P_i AND b

The first bound for P_1, found by Gilbert, MacWilliams and Sloane [10] using an uniform source distribution, is given by

$$P_1 \geq \frac{1}{\sqrt{b}}.$$

Afterwards this bound was proven under general conditions by Simmons and Brickell, they obtained

$$max\{P_0, P_1\} \geq \frac{1}{\sqrt{b}}.$$

We define a ***perfect authentication code*** as a code which achieves this bound. In [8] the following theorem is proved

Theorem 1 *There exists a perfect authentication code on r source states, $r \cdot k$ authenticated messages and k^2 encoding rules if and only if there exists a net of degree r and order k (see* [2], [5]*).*

The following combinatorial bound for P_i was proven by Schöbi [16] and Stinson [21]

Theorem 2 *In an authentication system there holds $P_i \geq (k-i)/(v-i)$, $i \geq 0$.*

Following Massey [14], an authentication system is *L-fold secure against spoofing* if

$$P_i = \frac{k-i}{v-i}, \quad \text{for all } i,\ 0 \leq i \leq L.$$

Stinson [21] proved an underbound for the number of encoding rules given in the next theorem

Theorem 3 *If a code achieves perfect L-fold secrecy and is $(L-1)$-fold secure against spoofing, then*

$$b \geq \binom{v}{L}.$$

An *optimal L-code* is defined as a code which has perfect L-fold secrecy, is $(L-!)$-fold secure against spoofing and which achieves the bound in the preceding theorem.

5. CONSTRUCTIONS OF CODES FOR ARBITRARY SOURCE DISTRIBUTION

5.1 AUTHENTICATION CODES DERIVED FROM GENERALIZED QUADRANGLES

Given an incidence structure P, in the sense of [5], we denote by $\Delta(P)$ the flag graph of P. Thus $\Delta(P)$ is the bipartite graph having vertex set the collection of points and blocks of P and edges the totality of flags (unordered incident point–block) pairs of P.

A *(thick) generalised polygon* ($n \in N$, $n \geq 2$) is an incidence structure P with the property that $\Delta(P)$ satisfies the following three conditions:

(i) each vertex has valency at least 3;

(ii) each pair of edges is contained in a circuit of length $2n$;

(iii) there is no circuit of length less than $2n$.

We deal with $n \geq 3$. Examples of generalised 3- and 4-gons are projective planes and generalised quadrangles which were used in [10] and [6] respectively to construct authentication schemes. The finite generalised n-gons, with which we shall be dealing exclusively, were classified by Feit and Higman [9]. These are tactical configurations (which means that every point is incident with the same number of lines) and we follow the convention of denoting the number of points on a line by $s+1$, $s>1$, and the number of lines through a point by $t+1$, $t>1$.

For a vertex X of Δ we denote by $\Delta_d(X) = \{Y \in \Delta \mid \delta(X,Y) = d\}$, i.e. the set of all vertices at distance d from X.

If there exists a finite generalised n-gon of order $n \geq 3$ with parameters s and t we can define, making use of an arbitrary point X, a Cartesian authentication scheme with

$$|S| = |\Delta_1(X)| = t+1 \tag{1}$$

$$|M| = |\Delta_d(X)| = \begin{cases} (t+1)t^{d-1} & d \text{ odd} \\ (t+1)s^{d/2}t^{d/2-1} & d \text{ even.} \end{cases} \tag{2}$$

and

$$|K| = |\Delta_n(X)| = \begin{cases} t^{n-1} & n \text{ odd} \\ s^{n/2}t^{n/2-1} & n \text{ even.} \end{cases} \tag{3}$$

The encoding rule e_Z determined by the key Z maps a source Y to

$$e_Z(Y) = Z_d$$

where Z_d is the vertex at distance d from X in the unique path $< Y, Z >$. A receiver with knowledge of the key Z authenticates Z_d by checking that it is at distance $n-d$ from Z. The corresponding source state Y is simply the vertex adjacent to X in the path $< X, Z_d >$. Starting with a line instead of a point results in an interchange of the parameters s and t in the above expression.

We give below a complete list of these Cartesian schemes which are 0-fold secure against spoofing, for all possible values of n. An entry * in the tables means that the value depends upon the structure of the generalised n-gon

$n = 3$: projective planes
$|S| = t+1$, $|K| = t^2$

d	2
$\|M\|$	$t(t+1)$
P_0	$1/t$
P_1	$1/t$
$\|E\|$	t^2

$n = 4$: generalised quadrangles
$|S| = t+1$, $|K| = s^2t$

d	2	3
$\|M\|$	$s(t+1)$	$st(t+1)$
P_0	$1/s$	$1/st$
P_1	$1/s$	$1/s$
$\|E\|$	*	s^2t

$n = 6$: generalised hexagon
$|S| = t+1$, $|K| = s^3t^2$

d	2	3	4	5
$\|M\|$	$s(t+1)$	$st(t+1)$	$s^2t(t+1)$	$(st)^2(t+1)$
P_0	$1/s$	$1/st$	$1/s^2t$	$1/(st)^2$
P_1	$1/s$	$1/st$	$1/st$	$1/s$
$\|E\|$	*	*	s^3t^2	s^3t^2

$n = 8$: generalised octagon
$|S| = t+1$, $|K| = s^4t^3$

d	2	3	4	5	6	7
$\|M\|$	$s(t+1)$	$st(t+1)$	$s^2t(t+1)$	$(st)^2(t+1)$	$s^3t^2(t+1)$	$(st)^3(t+1)$
P_0	$1/s$	$1/st$	$1/s^2t$	$1/(st)^2$	$1/s^3t^2$	$1/(st)^3$
P_1	$1/s$	$1/st$	$1/s^2t$	$1/s^2t$	$1/st$	$1/s$
$\|E\|$	*	*	*	s^4t^3	s^4t^3	s^4t^3

For a complete treatment of the security and implementation of these schemes we refer to [7], [8]

Using these incidence structures we can define perfect schemes in the following way. Let X, Y be two vertices at distance $2m$ of a $2m$-gon. We define the trace(X,Y) as the set

$$\text{trace}(X,Y) = \Delta_m(X) \cap \Delta_m(Y).$$

There holds

$$|\text{trace}(X,Y)| = \begin{cases} t+1, & \text{for } X,Y \text{ points} \\ s+1, & \text{for } X,Y \text{ lines} \end{cases}$$

The span of the pair (X,Y) is defined as

$$\text{span}(X,Y) = \{Z \mid Z \in \Delta_m(W),\ \forall W \in \text{trace}(X,Y)\}.$$

We have

$$|\text{span}(X,Y)| \leq \begin{cases} t+1 & \text{for } m \text{ even, } X,Y \text{ points} \\ & \text{for } m \text{ odd, } X,Y \text{ lines} \\ s+1 & \text{for } m \text{ even, } X,Y \text{ lines} \\ & \text{for } m \text{ odd, } X,Y \text{ points.} \end{cases}$$

The vertex X is called *regular* if and only if

$$|\text{span}(X,Y)| = \begin{cases} t+1 \\ s+1 \end{cases} ,\forall Y \in \Delta_{2m}(X).$$

Now consider the authentication scheme based on a regular vertex X with $d = m$. Then two keys Z and Z' produce the same encoding rule precisely when $Z' \in$ span(X,Z), or equivalently, when span(X,Z)=span(X,Z'). Hence the number of distinct encoding rules is given by:

$$|E| = \begin{cases} |K|/t, & m \text{ even} \\ |K|/s, & m \text{ odd.} \end{cases}$$

Thus, if X is regular and if we restrict our key set to one value from each span containing X, then for the authentication scheme with $d = m$, we have:

$$P_0 = P_1 = \begin{cases} 1/(st)^{(m-1)/2} & m \text{ odd} \\ 1/(s^{m/2}t^{m/2-1}) & m \text{ even} \end{cases}$$

and hence $P_0 = P_1 = 1/\sqrt{|E|}$. So the scheme is perfect in the sense of [19].

5.2 AUTHENTICATION CODES DERIVED FROM GENERALISED QUADRANGLES

Given a generalised quadrangle G of order (s,t), a *spread* is defined as a set of lines $\mathcal{R}$ such that each point of G is incident with exactly one line of $\mathcal{R}$. For further information on generalised quadrangles we refer to [15].

Consider again a GQ $\mathcal{G}$ of order (s,t) which contains a spread $\mathcal{R}=\{L_1,\ldots,L_{st+1}\}$. Define the source states as the lines of $\mathcal{R}$ ($k = st+1$) and the messages as the points of $\mathcal{G}$ ($v = (st+1)(s+1)$). Denote the points as $x_{1,1}, x_{1,2}, \ldots, x_{i,j}, \ldots, x_{st+1,s+1}$, with $x_{i,j}\ I\ L_i$, $1 \leq j \leq s+1$, $1 \leq i \leq st+1$.
Then we define an encoding rule in the following way. We associate with each point $x_{i,j}$ an encoding rule

$$e_{x_{i,j}}(L_k) \;=\; x_{i+k,l'},$$

with $x_{i+k,l'}$ the unique point on the line L_{i+k} which is collinear with $x_{i,j}$ (where $i+k$ is taken (mod $st+1$)). In this way we obtain $b = (1+s)(1+st)$ encoding rules.

Theorem 4 ([6]) *If there exists a GQ of order (s,t) containing a spread $\mathcal{R}$, then there is an optimal 1-code for $st+1$ source states and $(st+1)(s+1)$ messages.*

5.3 AUTHENTICATION CODES DERIVED FROM STEINER SYSTEMS

Consider a t-(v,k,λ) design $\mathcal{D}$. For $\lambda = 1$, these are the so called *Steiner systems* (see [1], [3], [12]).
In a t-$(v,k,1)$ design $\mathcal{D}$, each element occurs in $r = (v-1)\cdots(v-t+1)/(k-1)\cdots(k-t+1)$ blocks and the total number of blocks is given by $v\cdot(v-1)\cdots(v-t+1)/k\cdot(k-1)\cdots(k-t+1)$. We construct $k!$ encoding rules from every block of $\mathcal{D}$, since for each block $A = \{x_1,\ldots,x_k\}$ this is the number of keys required to do a perfect enciphering on the k points. Hence we obtain

$$b = \frac{v\cdot(v-1)\cdots(v-t+1)}{k\cdot(k-1)\cdots(k-t+1)}\cdot k! = \frac{v!(k-t)!}{(v-t)!}$$

encoding rules, which we shall use with probability $1/b$.

Theorem 5 ([6]) *A Steiner system $\mathcal{D}$ defines an $AC(k,v,v!(k-t)!/(v-t)!)$ which has perfect t-fold secrecy and $(t-1)$-fold security against spoofing.*

REMARK The foregoing construction can be applied to a more general structure, nl. a group-divisible t-design.
A *group-divisible t-design* $GD(k,\lambda,n,t,v)$ is a triple (X,G,A) satisfying:

1. X is a set of v elements called *points*
2. G is a partition of X into v/n subsets of n points, called *groups*
3. A is a set of subsets of X (called *blocks*), each of size k, such that a group and a block contain at most one common point
4. every t points of distinct groups occur in exactly λ blocks.

Note that a $GD(k,\lambda,n,t,k\cdot n)$ is equivalent with a *transversal t-design* (see [11]). Applying the same construction as in 6.3 a $GD(k,\lambda,n,t,v)$ defines an

$$\mathrm{AC}\left(\mathrm{k},\mathrm{v},\frac{\lambda\cdot\mathrm{v}\cdot(\mathrm{v}-\mathrm{n})\cdots(\mathrm{v}-(\mathrm{t}-1)\cdot\mathrm{n})}{\mathrm{k}\cdot(\mathrm{k}-1)\cdots(\mathrm{k}-\mathrm{t}+1)}\cdot\mathrm{k}!\right)$$

which has perfect t-fold secrecy and for which $P_{d_i}=(k-i)/(v-i\cdot n)$, for $0\le i\le t-1$. Moreover the code is $(t-1)$-fold secure against spoofing if and only if $n=1$, in which case we have a t-(v,k,λ) design.

6. AUTHENTICATION CODES FOR UNIFORM SOURCE DISTRIBUTION

We consider the construction of authentication codes for uniform source distributions ($p(s)=1/k$, for any source state s). As before we are dealing only with codes without splitting. We know that the best bound is given by $P_{d_i}=(k-i)/(v-i)$, for a spoofing attack of order i.

Theorem 6 *An authentication system is L-fold secure against spoofing w.r.t. the uniform probability distribution on the source states if and only if, for every i, $0\le i\le L$ and for every $M'\subset M$, $|M'|=i+1$,*

$$\sum_{e\in E(M')} p(e) = \frac{k}{v}\cdot\frac{k-1}{v-1}\cdots\frac{k-i}{v-i}.$$

REMARK In many authentication codes, the encoding strategy is to choose each encoding rule with probability $1/b$. If we assume that this encoding strategy is in fact optimal, then the properties of the foregoing theorem are of purely combinatorial nature. We can formulate the following theorem.

Theorem 7 *An authentication system is L-fold secure against spoofing with respect to a uniform encoding strategy and a uniform probability distribution on the source states if and only if the following property is valid for every i, $0 \leq i \leq L$ and every $M' \subset M$, $|M'| = i+1$,*

$$|E(M')| = b \cdot \frac{k}{v} \cdots \frac{k-i}{v-i}.$$

EXAMPLE *A t-(v,k,λ) design (see* [1], [3], [12]) *defines an authentication system for a uniform source distribution and a uniform encoding strategy $AC(k,v,b)$ which is $(t-1)$-fold secure against spoofing.*

REFERENCES

[1] Beth, T., Jungnickel, D., Lenz, H.: Design Theory, Wissenschaftsverlag Bibliografisches Institut Mannheim, 1985.

[2] Bose, R. C.: Graphs and designs, in: Finite geometric structures and their applications, ed. A. Barlotti, Ed. Cremonese Roma, 1973, 1–104.

[3] Cameron, P. J., Van Lint, J. H.: Graph Theory, Coding Theory and Block Designs, Lond. Math. Soc. Lect. Notes 19, Camb. Univ. Press, 1975.

[4] Brickell, E. F.: A few results in message authentication, in: Proc. of the 15th Southeastern Conf. on Combinatorics, Graph theory and Computing, Boca Raton LA, 1984, 141–154.

[5] Dembowski, P.: Finite Geometries, Springer Verlag, 1968.

[6] De Soete, M.: Some Constructions for Authentication / Secrecy Codes, in: Advances in Cryptology–Proceedings of Eurocrypt '88, Lect. Notes Comp. Science 330, Springer 1988, 57–75.

[7] De Soete, M., Vedder, K. and Walker, M.: Authentication Schemes derived from Genersalised Polygons, in: Advances in Cryptology–Proceedings Eurocrypt '89, Lect. Notes Comp. Science, to appear.

[8] De Soete, M. , Vedder, K., and Walker, M.: Cartesian Authentication Codes, in preparation.

[9] Feit, W. and Higman, G.: The non–existence of certain generalised polygons, J. Algebra 1 (1964), 114–131.

[10] Gilbert, E. N., MacWilliams, F. J. and Sloane, N. J. A.: Codes which detect deception, Bell Sys. Techn. J., Vol. 53-3 (1974), 405–424.

[11] Hanani, H.: A Class of Three-Designs, J.C.T.(A) 26 (1979), 1–19.

[12] Hughes, D. R. and Piper, F. C.: Design theory, Cambridge University Press, 1985.

[13] Jungnickel, D.: Graphen, Netzwerke und Algorithmen, Wissenschaftsverlag Bib. Inst. Zürich, 1987.

[14] Massey, J. L.: Cryptography - A Selective Survey, in: Proc. of 1985 Int. Tirrenia Workshop on Digital Communications, Tirrenia, Italy, 1985, Digital Communications, ed. E. Biglieri and G. Prati, Elsevier Science Publ., 1986, 3–25.

[15] Payne, S. E. and Thas, J. A.: Finite generalized quadrangles, Research Notes in Math. #110, Pitman Publ. Inc. 1984.

[16] Schöbi, P.: Perfect authentication systems for data sources with arbitrary statistics, Eurocrypt 1986, Preprint.

[17] Shannon, C. E.: Communication Theory of Secrecy Systems, Bell Technical Journal, Vol. 28 (1949), 656–715.

[18] Simmons, G. J.: Message Authentication: A Game on Hypergraphs, in: Proc. of the 15th Southeastern Conf. on Combinatorics, Graph Theory and Computing, Baton Rouge LA Mar 5–8 1984, Cong. Num. 45, 1984, 161- 192.

[19] Simmons, G. J.: Authentication theory / Coding theory, in: Proc. of Crypto'84, Santa Barbara, CA, Aug 19–22, 1984, Advances in Cryptology, ed. R. Blakley, Lect. Notes Comp. Science 196, Springer 1985, 411–432.

[20] Simmons, G. J.: A natural taxonomy for digital information authentication schemes, in: Proc. of Crypto '87, Santa Barbara, CA, Aug 16–20, 1987, Advances in Cryptology, ed. C. Pomerance, Lect. Notes Comp. Science 293, Springer 1988, 269–288.

[21] Stinson, D. R.: A construction for authentication / secrecy codes from certain combinatorial designs, J. of Cryptology Vol. 1, nr. 2 (1988), 119–127.

[22] Tits, J.: Classification of buildings of spherical type and Moufang polygons, Atti dei Convegni Lincei 17 (1976), 230–246.

GEOMETRIC THRESHOLD SCHEMES

M. De Soete
MBLE-I.S.G., Brussel, Belgium

ABSTRACT

This paper gives constructions of infinite classes of 2, 3 and 4-threshold schemes based on finite incidence structures such as generalised quadrangles and projective planes.

1. INTRODUCTION

Any scheme which is to protect information has to be designed with the following three main points in mind: possible loss or destruction of the information or parts thereof, attack from inside or outside to obtain or destroy the information and efficiency. One obvious way to guard the information against loss or destruction is to make multiple copies of it and distribute them amongst trustworthy parties. This has two obvious drawbacks. Too few copies might cause the loss of the information while too many copies could lead to the information falling into wrong hands. Moreover, each trusted party is in the possession of all of the information.

In 1979 Blakley and Shamir independently introduced what is known under the name "threshold schemes". In those schemes pieces of information are distributed amongst "trustees" in such a way that any number of trustees which achieve a quorum or threshold can reconstruct the information. Clearly "reconstruction of the information" can be replaced by "gaining access", "starting a computer program" or anything which is similar to this. A more formal definition reads as follows.

A *t-threshold scheme* consists of $s \geq t$ pieces of information (*shadows*), such that

(i) a *secret datum* X can be retrieved from any t of the s shadows

(ii) X cannot be determined from any $t-1$ or fewer of the s shadows.

The second condition needs some explanation. First of all, it means that the knowledge of $t-1$ shadows should suggest every possible datum with about the same probability. If the number of possible data is finite, then one can, of course, guess the correct datum in a finite amount of time and the knowledge of $t-1$ shadows might even reduce the time necessary. It should, however, be beyond any reasonable computing time.

The security considerations depend on the nature of the secret datum X. If the value of X is, for instance, the master key of a cryptosystem ([3], [9]), then a correct guess of X compromises the system. The probability to do this might be different to the probability to cheat the system by entering "made-up" shadows. If the knowledge of X is by itself of no use, X might be a trigger to start a computer program, then this probability determines the security level. The possible difference of these two probabilities is illustrated by the schemes given in section 3.3.

In the above definition the number s stands for the maximum number of shadows one can hand out to the trustees. If $s = t$, the loss of any one shadow is, by definition, equivalent to the loss of the secret datum. This is also the case, if $s > t$ but the number of shadows handed out is equal to t. Administrative procedures such as a back-up list of all shadows, of course, prevent such a break down but impair the security. Hence it is advantageous to the designer of a t-threshold scheme, if he has some room of manoeuvre between t and s. This allows him to fix the number of distributed shadows according to his needs.

We will discuss classes of threshold schemes with $t = 2, 3$ (see [4]) and 4 (see [2]) which have the property that the level of security and with it the number s can be chosen as high and large as desired. They are based on finite generalised quadrangles and finite projective spaces. The quadrangles also allow the construction of threshold schemes which cater for the situation where the trustees do not trust each other and a threshold has to be achieved in each one of a number of distrusting parties. This could, for instance, also be useful in a situation which involves not only human beings but say computer programs as well. We conclude this introduction with a definition of such threshold schemes.

A $(t_1, \ldots, t_n)$-*threshold scheme* is a t-threshold scheme with $t = \sum_{i=1}^n t_i$ where the set of shadows is partitioned into n subsets B_i $(i = 1, \ldots, n)$, with $|B_i| = s_i$, $\sum_{i=1}^n s_i = s$, and a quorum of $t_i \leq s_i$ is needed in each set B_i. If just n thresholds $t_1, \ldots, t_n$ have to be achieved and it does not matter in which one of the sets B_i, we call it a $(t_1, \ldots, t_n)^*$-threshold scheme.

2. GEOMETRIC BACKGROUND

A (finite) *generalised quadrangle* (GQ) of *order* (σ, τ) is an incidence structure (P, B, I) which consists of two non–empty sets P and B and a subset $I \subset P \times B$. The elements of P and B are called points and lines resp. and I is a symmetric incidence relation which satisfies the following axioms:

1. Each point is incident with exactly $1 + \tau$ lines $(\tau \geq 1)$ and two distinct points are incident with at most one line.

2. Each line is incident with exactly $1 + \sigma$ points $(\sigma \geq 1)$ and two distict lines are incident with at most one point.

3. For every point x and every line L which are not incident with each other, there exists a unique line which is incident with both x and a (unique) point on L.

It follows from this definition that every GQ of order (σ, τ) has associated with it a GQ of order (τ, σ) which is obtained by interchanging the rôles of the points and lines. We call it *the dual GQ*. This implies that in any definition or theorem the words "points" and "lines" and the parameters "σ" and "τ" may be interchanged.

The definition allows us to identify each line with the set of points it is incident with. The set of points collinear with a point x is denoted by $x^\perp$ (note that $x \in x^\perp$).

The proof of the following lemma is left to the reader as an easy exercise with the exception of (iii) a proof of which can be found in [8].

Lemma 1 *Let (P, B, I) be a generalised quadrangle of order (σ, τ), then*

(i) $|P| = (\sigma + 1)(\sigma\tau + 1)$, $|B| = (\tau + 1)(\sigma\tau + 1)$

(ii) $|x^\perp| = 1 + (\tau + 1)\sigma$ for all points $x \in P$

(iii) $\sigma + \tau$ divides $\sigma\tau(\sigma + 1)(\tau + 1)$.

The threshold schemes we are going to introduce are based on *the span* of pointsets. The *trace* of a pair (x, y) of distinct points is defined to be the set $x^\perp \cap y^\perp$ and is denoted as tr$(x,y)=\{x,y\}^\perp$. More generally, one can define for $A \subset P$, the set $A^\perp = \cap \{x^\perp \mid x \in A\}$. The span of two distinct points x and y, is defined as sp$(x,y)=\{x,y\}^{\perp\perp}=\{u \in P \mid u \in z^\perp \ \forall z \in \text{tr}(x,y)\}$. Hence it consists of all points which are collinear with every point in the trace of x and y.
If x and y are not collinear, we note that x, y are in sp(x, y), no two points of sp(x, y) are collinear and $|\text{sp}(x,y)| \leq \tau + 1$. The latter follows since the points of sp(x, y) have to be contained in the $\tau + 1$ lines through any of the points of $x^\perp \cap y^\perp$.

Finally, a *triad* (of points) is a triple of mutually non-collinear points' Given a triad $T = (x, y, z)$, a *centre* of T is just a point of $T^\perp$ =tr(x, y, z).
The reader who is interested in finding out more about the theory of generalised quadrangles is referred to the book by Payne and Thas [8].

The 4-threshold schemes are based on Baer subplanes of finite projective planes. A subplane Π' of a finite projective plane $\Pi = PG(2,q)$ consists of subsets of the sets of points and lines of Π such that Π' is itself a projective plane. By a well known theorem of Bruck (see [6]), there holds that the order q' of a subplane of $PG(2, q)$ satisfies $q'^2 = q$ or $q'^2 + q' \leq q$. A Desarguesian plane [6] of order $q = p^h$ contains a subplane for every prime power $q' = p^{h'}$, where h' is a divisor of h. Conversely, the order of every subplane is of this form. Subplanes of order $q' = \sqrt{q}$ are called *Baer subplanes*. They have the important property that every line of Π, which is not a line of Π', is a tangent to the subplane (i.e. contains exactly one point of Π') and dually, every point of Π, which is not a point of the subplane, is on exactly one line of the Baer subplane Π'.

An *arc* of a finite projective plane is a set of points no three of which are collinear. A maximal arc (which means an arc containing the maximum number of points possible) is called an *oval*. The number of points on an oval equals $q + 1$ for q odd and $q + 2$ for q even. If the arc only contains four points, we call it a *quadrangle*. It is easily seen that every projective plane contains a quadrangle.

3. THE SCHEMES BASED ON GENERALISED QUADRANGLES

3.1 THE 2-THRESHOLD SCHEMES

Let G be a generalised quadrangle of order (σ, τ) with $\sigma, \tau > 1$, and let x and y be two non-collinear points of G. Then the points of $\mathrm{sp}(x, y)$ can be used as the shadows of a 2-threshold scheme with the secret datum X being the span of x and y.

For consider two distinct points w and z of $\mathrm{sp}(x, y)$. As points of the span they are not collinear but each one of them is collinear with every point in $x^\perp \cap y^\perp$. Hence $z^\perp \cap w^\perp = x^\perp \cap y^\perp$ and $\mathrm{sp}(z, w) = \mathrm{sp}(x, y) = X$. So the secret datum is determined by any two of the shadows.

The probability to obtain X with the knowledge of no or just one shadow depends on the number of shadows in X. This number is subject to the structure of G and the particular choice of the span. It is however, never greater than $\tau + 1$. We obtain the following expression for the possibility that the secret datum is revealed by entering a valid shadow and some other point.

$$\mathrm{Prob} = \frac{\mathrm{s} - 1}{\sigma^2\tau + \sigma\tau + \sigma} \leq \frac{\tau}{\sigma^2\tau + \sigma\tau + \sigma} \cdot \quad (3.1)$$

When setting the security level one has, however, to take into account that a trustee knows some finite geometry and for some reason or other the lines through his own shadow. This increases his probability of a successful attempt to break the system to

$$\mathrm{Prob} = \frac{\mathrm{s} - 1}{\sigma^2\tau + \sigma\tau + \sigma - (\sigma\tau + \sigma)} = \frac{\mathrm{s} - 1}{\sigma^2\tau} \leq \frac{1}{\sigma^2} \quad (3.2)$$

as he can rule out the $\sigma\tau + \sigma$ points which are collinear with his shadow. Equation (3.2) implies that the security level only depends on σ or, in other words, the number of points on a line, if $\mathrm{sp}(x, y)$ contains $\tau + 1$ points. If this is the case, the pair (x, y) is called *regular*. A point x is said to be regular, if for every y, $y \not\sim x$, the pair (x, y) is regular.

So far we have not said anything about the existence of generalised quadrangles. If a point of a GQ is regular then $\sigma \geq \tau$ (see [8]). So the smallest case is $\sigma = \tau$. Such generalised quadrangles exist indeed. The ones in which all the points are regular are derived from the projective geometry $PG(3,q)$. The points of the GQ are just the points of $PG(3,q)$ while the lines are the totally isotropic lines with respect to a symplectic polarity. For the necessary background in finite geometry the reader is referred to [1], [7]. As these geometries exist for every prime power q, we have obtained an infinite class of 2-threshold schemes which admit $q+1$ shadows at a security level of $1/q^2$ and have an implementation size of q^3+q^2+q+1 points and lines. Since these generalised quadrangles are coordinatised (see [8]), they can be implemented on a computer.

Using a regular pair of points for an implementation supplies us with at least $\tau+1 \geq \sqrt{\sigma}+1$ shadows at a security level of $1/\sigma^2$ since the inequalities $\tau^2 \geq \sigma \geq \tau$ hold (see [8]). Such a number is in nearly all cases far beyond anything needed. So the question arises whether one should use a non-regular pair of points whose span is sufficiently large. A span containing s points increases the security level to $(s-1)/\tau\sigma^2$ at the same order (σ,τ). For instance, the generalised quadrangles derived from a non-singular hermitian variety in $PG(4,q^2)$ have order (q^2,q^3). Here the spans consist of $q+1$ points. Hence the probability to cheat is approximately $1/q^6$ while the above examples attain a security level of only $1/q^4$ at the same linesize. This is, however, not the only criterion for the magnitude of the implementation.

It should be mentioned that regular pairs have a non-negligible advantage when it comes to the actual implementation, since we can make use of the following observation. Two points x' and y' belong to $\mathrm{sp}(x,y)$ if and only if they are collinear with every one of the points in $x^\perp \cap y^\perp$. Checking this is clearly not feasible. If the pair (x,y) is regular, it suffices to show that x' and y' are collinear with just two of those points. Since in this case the trace of a span is equal to the span of the trace. So we just have to store two points of the trace and check whether x' and y' are collinear with both of them. The amount of computation needed for this depends on the number of coordinates and the particular field used for the coordinatisation.

3.2 THE 3-THRESHOLD SCHEMES

The threshold schemes constructed in the preceding section were based on pairs of non-collinear points. Now we are going to use triads of points. We will see that, when assessing the security of the new systems, it is not sufficient to just transfer the considerations made for the 2-threshold schemes. The "extension" will provide an attacker with new possibilities.

Let (x, y, z) form a triad, and let $\mathrm{sp}(x, y, z) = \{x, y, z\}^{\perp\perp}$ be the secret datum X. It is easy to see that any three points of X uniquely determine X. So condition (i) for a 3-threshold scheme is satisfied.
Two disloyal trustees with respective shadows x', y' have a success rate of

$$(s-2)/(\sigma^2\tau+\sigma\tau+\sigma-1) \quad (3.3)$$

in a staight forward attack. If they can rule out the $2\sigma(\tau+1)-(\tau+1) = 2\sigma\tau+2\sigma-\tau-1$ points which are collinear with x', y', then their probability to break the system is

$$\mathrm{Prob} = \frac{s-2}{\sigma^2\tau-\sigma\tau-\sigma+\tau}. \quad (3.4)$$

So far everything is similar to the case of two non-collinear points. Being abl to rule out the points of $\mathrm{tr}(x', y')$, however, opens up new ways of breaking the system in this situation as we will see later.

The number of shadows depends on the underlying GQ. If this is of order (σ, σ^2) with $\sigma > 1$, then $\mathrm{tr}(x, y, z) = \{x, y, z\}^{\perp}$ always consists of $\sigma + 1$ points and hence $\mathrm{sp}(x, y, z)$ contains at most $\sigma+1$ points. The point x is 3-regular, if $|\mathrm{sp}(x, y, z)| = \sigma+1$ for any triad (x, y, z) through x in G. Hence X contains $s = \sigma + 1$ shadows.

Examples of such generalised quadrangles are $Q(5, q)$, the elliptic quadrics in $PG(5, q)$, for every prime power q. These give rise to 3-threshold schemes with $q + 1$ shadows. We will discuss the security using the generalised quadrangles of order (σ, σ^2). For these Equation (3.4) reads

$$\mathrm{Prob} = \frac{\sigma-1}{\sigma^4-\sigma^3+\sigma^2-\sigma} = \frac{1}{\sigma^3+\sigma}. \quad (3.5)$$

If the two trustees x' and y' can work out the points of $\mathrm{tr}(x', y')$ they could make use of this knowledge and the relationship between a trace and its span. They take any point u in $\mathrm{tr}(x, y)$, choose a line L through this point and a point $g \neq u$ on L. The probability that u is in $\mathrm{tr}(x, y, z)$ is $(\sigma + 1)/(\sigma^2 + 1)$, the one for L to intersect sp(x, y, z) in a point different to x and y is $(\sigma - 1)/(\sigma^2 - 1)$, while the probability that g is indeed this point is $1/\sigma$. Assuming that the three events are independent the two disloyal trustees succeed in breaking the system with a probability of

$$\frac{\sigma+1}{\sigma^2+1} \cdot \frac{\sigma-1}{\sigma^2-1} \cdot \frac{1}{\sigma} = \frac{1}{\sigma^3+\sigma}. \quad (3.6)$$

So all this effort has not increased their chances. An improvement of this attack can be made if one knows conditions under which a line L through z does or does not intersect sp(x, y, z) and the checking of these conditions could be done without the system knowing it. Being able to determine a correct line raises the "success rate" to $(\sigma + 1)/(\sigma^3 + \sigma)$.

Clearly a lot of computing would have to go into such an attack. Any decrease in the security level given by (3.4) was based on the assumption that the trustees know not only their coordinates but also enough about the implementation to work out tr(x', y'). If they can do this it is also fair to assume that they can determine a point of sp(x', y') and feed the system this point. As sp(x, y, z) is contained in sp(x', y') the security now depends only on the size of sp(x', y') which is bounded above by $\sigma^2 + 1$. This yields a probability of

$$\text{Prob} = \frac{\sigma-1}{|\text{sp}(x', y')|} \geq \frac{\sigma-1}{\sigma^2-1} = \frac{1}{\sigma+1}. \quad (3.7)$$

Hence, if the trustees know the underlying implementation, the security level depends only on the span of x' and y' and might be unacceptable.

There is clearly no need for a trustee to know "his" shadow but one cannot rule out the possibility that he does. There is, however, in this scheme a way to prevent the trustee from making use of his knowledge. Before the system checks the shadows for their validity it does apply a secret coordinate transformation to them. So the secret datum X is not the span of the points x,y and z but of their transforms. This renders the knowledge of both tr(x', y') and sp(x', y') a useless information and increases the security level to the security level given in (3.4).

3.3 COMBINED SCHEMES

Distinct threshold schemes defined on the same underlying GQ obviously give rise to $(t_1, \ldots, t_n)$-threshold schemes. Using the geometry of the GQ allows the construction of more sophisticated schemes.

Let G be a generalised quadrangle with $\sigma > \tau$ in which every point is regular. To construct a $(1,2)^*$-threshold scheme we choose a triad (x, y, z) where z is not collinear with any point in $\mathrm{sp}(x, y)$. The condition $\sigma > \tau$ guarantees the existence of such triads since there are $\tau(\sigma - \tau)(\sigma - 1)$ points z for every pair (x, y) of non-collinear points. As the secret datum X we select an arbitrary point of $\mathrm{tr}(x, y)$. Putting $B_1 = \mathrm{sp}(\mathrm{x}, \mathrm{y})$ and $B_2 = \mathrm{tr}(\mathrm{X}, \mathrm{z})$ we obtain a $(1,2)^*$-threshold scheme.

To verify this we note that z is not collinear with X as (x, y) is a regular pair. The regularity of all points also implies that every triad has exactly 0,1, or $\tau + 1$ centres (see [8]).

Let x', y' be two shadows of B_1 and z' a shadow of B_2. If they form a triad, then, in view of Axiom(iii), X is the unique centre of this triad. If z' and, say x' are collinear, then X is the unique point on the line through z' and x' which is collinear with y'. Now consider the case that two shadows are in B_2 and one is in B_1. The trace T of the two points in B_2 has exactly one point in common with $\mathrm{tr}(x, y)$, namely the point X. This is the only point of T which is collinear with the shadow in B_1.

Two non-collinear shadows, whether or not they belong to the same class, determine a trace which contains X and τ further points. Hence their probability to guess X is

$$\frac{1}{\tau + 1} \quad (3.8).$$

Even if all the trustees of one class join their forces they cannot improve this probability. If the two shadows are collinear, then X is one of the $\sigma - 1 \geq \tau$ points on their common line. So this case gives a probability of

$$\frac{1}{\sigma - 1} \leq \frac{1}{\tau} \quad (3.9).$$

We note that there are no non-trivial examples known of generalised quadrangles with $\sigma = \tau + 1$. Examples which can be used are the duals of those mentioned in the preceding section. They are of order (q^2, q), where q is any prime power.

Using the same kind of implementation as before one can check that the shadows belong to the correct classes. We store three points X, z and w, where w is in $\mathrm{tr}(x, y)$. When three points together with their respective "class numbers" are entered, the system checks that they are collinear with the So we have joined two 2-threshold schemes to form a $(1,2)^*$-threshold scheme.

Since the system checks the entered values for the correct class, the probability to break the system is smaller then the ones given above, if the knowledge of X in itself is not equivalent to a compromise of the system.
There are several ways to construct a possible third shadow. None of these yields a better probability than trying to figure out X first and then a "correct" shadow. So the probability in (3.8) has to be multiplied by $1/(\sigma - 1)$ and the one given in (3.9) by $1/\sigma$. So the chances to enter a correct third shadow are about $1/\tau^2$.
It should be mentioned that a coordinate transformation will reduce all these probabilities to about 1 over the number of points of the GQ. So two trustees stand no better chance than two outsiders who just know the underlying GQ.

We conclude this section with an example involving a "supershadow". Let (x, y, z) be a triad such that z is not in $\mathrm{sp}(x, y)$. Then $\mathrm{sp}(x, y)$ and $\mathrm{sp}(x, z)$ have just the point x in common. We define three classes $B_1 = \{x\}$, $B_2 = \mathrm{sp(x,y)}\backslash\{\mathrm{x}\}$ and $B_3 = \mathrm{sp(x,z)}\backslash\{\mathrm{x}\}$, and let $X = \mathrm{tr(x,y)} \cup \mathrm{tr(x,z)}$. This yields both a (1,1,1)- and a (0,2,2)-threshold scheme with the shadow x being more powerful than the other shadows. We note that $\mathrm{tr}(x, y)$ and $\mathrm{tr}(x, z)$ intersect in a unique point u, say. So, if every point is regular, we only need to store u and a further point in each trace. We leave it to the reader work out the various probabilities to cheat the system.

4. THE 4-THRESHOLD SCHEMES BASED ON BAER SUBPLANES

Desarguesian planes whose orders are proper prime powers yield 4-threshold schemes in a natural way, since there is a unique subplane of a given admissible order through any quadrangle and there are many such subplanes through any three non-collinear points.

Let Π be the Desarguesian plane of order q. We choose a subplane Π' of order q'

of Π and in this plane $s \geq 4$ points in such a way that no three of them are collinear. Any four of the s shadows determine the subplane uniquely. The maximal number of shadows we can select depends only on the order m of Π', as the shadows form an arc in the subplane. This number is $q' + 2$ or $q' + 1$ according to q' even or odd. Likewise does the security depend on the order of the subplane.

Since we want to store as few information as possible about the subplane itself, we select a point X of Π' and choose a tangent S to Π' at the point X. There are $q + 1 - (q' + 1)$ such tangents, for any line through X, which is not a line of the subplane, is a tangent. So the system only knows X, and possibly, S. Whenever $u \geq 4$ shadows enter the system, it tries to construct a subplane of a given order through these points and, if there is a unique one, checks whether S is a tangent at X. As an additional feature the system could examine whether any three of the u points are collinear.
Whether we keep S secret depends on the particular implementation. If the subplane is a Baer subplane, then the probability that three shadows manage to break the system is in either case in the order of $1/q$. Only if q' is relatively small in comparison with q it makes sense to store S secretly. For a complete description of the possible attacks and the implementation of the system we refer to [2].

In a similar way one can construct t-threshold schemes for $t \geq 4$ by using Baer subspaces of the finite projective space $\mathcal{P}=PG(d,q)$, of dimension d and order q ([1],[7]). It is well known that for $d \geq 3$ every projective space can be constructed as follows. Let V be the vector space of dimension $d+1$ over the finite field with q elements. Then the points of $\mathcal{P}$ are the 1-dimensional subspaces of V, the lines of $\mathcal{P}$ are the 2-dimensional subspaces of V, etc..., while the hyperplanes of $\mathcal{P}$ are the d-dimensional subspaces of V. In other words, every projective space of dimension at least 3 is Desarguesian. We call a subspace of $\mathcal{P}$ of the same dimansion as $\mathcal{P}$ and order $\sqrt{q}$ a *Baer subspace.*
The constuction of $(d+2)$-threshold schemes is based on the following lemma.

Lemma 2 ([10], [2]) *Let $\mathcal{P}$ be the projective space of dimension d and order q. Then there is precisely one Baer subspace of $\mathcal{P}$ through any $d+2$ points of $\mathcal{P}$ no $d+1$ of which being in a common hyperplane.*

Using this lemma, $(d+2)$-threshold schemes can be constructed for every $d \geq 2$ in a similar way as for Desarguesian planes.

REFERENCES

[1] Beth, T., Jungnickel, D. and Lenz, H.: Design Theory, Wissenschaftsverlag Bibliographisches Institut Mannheim, 1985.

[2] Beutelspacher, A. and Vedder, K.: Geometric Structures as Threshold Schemes, in: The Institute of Mathematics and its Applications, Conf. Series 20, Cryptography and Coding, ed. H. J. Beker and F. C. Piper, 1989, 255–268.

[3] Blakley, G. R.: Safeguarding cryptographic keys, in: Proceedings NCC, AFIPS Press, Montvale, N.J., Vol. 48, 1979, 313-317.

[4] De Soete, M. and Thas, J.A.: A coordinatisation of the generalised quadrangles of order $(s, s+2)$, J. Comb. Theory A, Vol. 48-1 (1988), 1–11.

[5] Hanssens, G. and Van Maldeghem, H.: Coordinatisation of Generalised Quadrangles, Annals of Discr. Math. 37 (1988), 195–208.

[6] Hughes, D. R. and Piper, F. C.: Projective Planes, Springer Verlag, Berlin–Heidelberg–New York, 1973.

[7] Hughes, D. R. and Piper, F. C.: Design Theory, Cambridge University Press, 1985.

[8] Payne, S. E. and Thas, J. A.: Finite generalised quadrangles, Research Notes in Math. #110, Pitman Publ. Inc., 1984.

[9] Shamir, A.: How to share a secret, Communications ACM, Vol. 22 nr.11 (1979), 612–613.

[10] Sved, M.: Baer subspaces in the n dimensional projective space, Comb. Math. Proc. Adelaide (1982), 375–391.

W*M's: A SURVEY WRITING ON SOME BINARY MEMORIES WITH CONSTRAINTS

G. D. Cohen
Télécom Paris and CNRS URA 820, Paris, France

ABSTRACT

We survey some constructions of codes and estimations of capacities for binary memories subject to constraints on transitions between states. The capacity is found for the special class of translation-invariant constraints.

A summarizing table is given for write-unidirectional memories, write-isolated memories, and writing on reluctant, defective or suspicious memories.

I- INTRODUCTION : W*M's

We consider a binary storage medium consisting of n cells on which we want to store and update information. These operations must be performed under some constraints, dictated by technology, cost, efficiency, speed,... In the past few years, many models have been studied, which we list here in more or less chronological order : write-once memory or WOM ([18], [22], [5]), write-unidirectional memory or WUM ([3], [6], [20]) write-efficient memory or WEM [1] and write-isolated memory or WIM [7], [11].

We also consider the related problems of reluctant memories, WRM [9] and defective memories WDM [10], [14]. The reason for this blossoming are evoluting technology, fashion... and existing letters !

The initial model (WOM) representing the first generation of optical disks differs fondamentally from the others in that writing is irreversible. It has been much studied already and will almost not be considered here. Thus in this paper, * stands for U, E, I, R or D. Accent will be put on general methods rather than specific constructions and some already published proofs will be only sketched.

We assume we have a W*M with n positions, which we use for writing one message among M possible. We want to be able to continue the process indefinitely, under a specific constraint depending on *. The problem is : what is asymptotically the maximum achievable rate R of the W*M, defined as (log being to the base 2)

$$R := (1/n) \log M \qquad ?$$

Four cases can be distinguished, arising when :
The encoder (writer) and/or the decoder (reader) are informed or uninformed about the previous state of the memory.

We shall use the following notation :

Case 1 means encoder and decoder informed about the previous state.
Case 2 means encoder informed, decoder uninformed.
Case 3 means encoder uninformed, decoder informed.
Case 4 means encoder and decoder uninformed.

We shall investigate the W*M under two different hypotheses :

1) ε-*error* : errors can occur when decoding with arbitrarily small probability.

2) 0-*error* : no error occurs.
The maximum achievable rates or *capacities* will be denoted by C_ε and C_0 respectively.

With obvious notations, we thus have 8 capacities to consider :

$$C_{i,j}, \quad \text{with } i = 0, \ \varepsilon \ \text{ and } j = 1, 2, 3, 4.$$

Of course we have :

$$C_{\varepsilon,j} \geq C_{0,j}$$
$$C_{i,1} \geq C_{i,2}, \quad C_{i,3} \geq C_{i,4}.$$

Basic knowledge in Information Theory and Coding Theory will be assumed (see [8] and [16] respectively).

2. THE CONSTRAINTS

2.1. WUM

This model has bee introduced by Borden [3], whom we quote :

"A write-unidirectional memory is a binary storage medium which is constrained, during the updating of the information stored "to either writing 1's in selected bit positions or 0's in selected bit positions and is not permitted to write combinations of 0's and 1's. Such a constraint arises when the mechanism that chooses to write 0's or 1's operates much more slowly than the means of accessing and scanning a word".

2.2. WIM

A write-isolated memory is a binary storage medium on which no change of two consecutive positions is allowed when updating. This constraint is dictated by the nowaday technology for writing on some digital disks [21] and is studied in [7].

2.3. WRM

Suppose our storage media allows only a limited number, say r, of bit changes when updating. This question, refered to as writing on a reluctant memory, has been considered first by Fellows [9].

2.4. WDM

The problem of writing on a memory with defects has been considered by many authors (see [10] and [14]). The model is the following : a set S_0 of positions of the memory are stuck at "0", a set S_1 at "1". The sets S_0, S_1, are known to the encoder only and satisfy $|S_0|+|S_1|=s$.

2.5. WEM

This a general model, containing as special cases WOM, WUM, WIM and WRM, where costs are associated to transitions. It is introduced in [1], where the maximal rate achievable with a maximal cost per letter criteria is investigated.

3. A GENERAL UPPER BOUND ON CAPACITY IN THE ZERO-ERROR CASE

Let us give an upper bound, valid in all four cases. Let V_n be the set of binary n-tuples, $V_n(t)$ the set of possible states of the memory after t utilizations. Consider the following directed graph

$G = (V,E)$ with $V = V_n(t) \cup V_n(t+1)$ and
$E = \{(i, j),$ where $i \in V_n(t), j \in V_n(t+1)$ and $i \rightarrow j$ is allowed$\}$.
Now the following is clear :

<u>Proposition 1</u> $\quad M \leq \operatorname{Max}_i \upsilon(i), \quad i \in V_n(t), \quad$ (GB)

where $\upsilon(i)$ is the valency of i (indeed any state i can be updated to at most $\upsilon(i)$ states j).□

Specific Examples of (GB)

For WIM, WRM and WDM, it is easily seen that the valency of the graph G is independent of the state, so we may suppose we are in the state $\underline{0}$.

WIM : $\upsilon(\underline{0}) = |\{x_1, x_2,..., x_n) \in V_n : x_j\, x_{j+1} = 0 \text{ for } i = 0, 1,..., n-1\}|$.

It is well known that $\upsilon(0)$ is ϕ_n the n^{-th} Fibonacci number with $\phi_0 = 1$, $\phi_1 = 2$. Asymptotically

$$n^{-1} \log \phi_n \cong \log ((1+\sqrt{5})/2) \cong 0.69$$

That is $\mathbf{C}_{0,1} \leq 0.69$.

WRM : In that case

$$\upsilon(\underline{0}) = |\{\underline{x} \in V_n : w(\underline{x}) \leq r\}|$$

$$= \sum_{j=0}^{r} \binom{n}{j},$$

where w(x) is the Hamming weight of x, i.e. the number of 1's in x. If only a fraction $r = \lambda n$, $0 \leq \lambda \leq 1/2$, of the total number n of positions may change when updating, then

$$\upsilon(\underline{0}) = \sum_{j=0}^{\lambda n} \binom{n}{j} \cong 2^{nH(\lambda)}$$

where $H(\lambda) := -\lambda \log \lambda - (1-\lambda) \log (1-\lambda)$ is the entropy function. This gives as an upperbound on the capacity

$$C_{0,1} \leq H(\lambda).$$

For a WUM, $\upsilon(i)$ depends on the weight w of the state i :

$$\upsilon(i) = 2^w + 2^{n-w} - 1.$$

WDM : Here changes are impossible on S_oUS_1, the set of "stuck at" positions. Hence

$$\upsilon(\underline{0}) = 2^{n-s}$$

and $$C_{0,1} \leq 1 - {}^{s}/_{n} .$$

4. TRANSLATION-INVARIANT CONSTRAINTS.
EXACT FORMULAS FOR CAPACITY IN CASES 1 AN 2 (ZERO-ERROR).

Let us denote by F(x) the set of possible states for updating a given x in V_n, with $|F(x)| = \upsilon(x)$.

We shall say that the constraints are translation-invariant if

$$F(x) = x + F(\underline{0}) : = \{x+y : y \in F(\underline{0})\}.$$

We set $F(\underline{0}) = F$, $|F| = f_n$ and call F(x) the F-set centered at x.

Examples :

WIM. $F = \{x = (x_1, x_2, \ldots x_n\} \in V_n : x_j\, x_{j+1} = 0 \text{ for } i = 0, 1, \ldots, n-1\}$

WRM. $F = \{x \in V_n : w(x) \leq r\}$.

Here F is the Hamming sphere of radius r centered at $\underline{0}$.

WDM. $F = \{x \in V_n : x_i = 0 \text{ for } i \in S_0\, US_1\}$.

That is, F is a cylinder of radius n-s and length s.

For the remaining of this paragraph, we assume we are in case 2 : *Encoder knows the previous state of the memory (i.e can read before writing) but decoder does not.*

We now present a coding strategy based on the notion of good blocks, which is a straightforward extension of the treatment presented in [7] for a WIM and will apply whenever the constraints are translation-invariant.

A subset $B = \{b_1, b_2, \ldots b_m\}$ of V_n is called a good block if :

$$\bigcup_{b \in B} F(b) = V_n . \qquad (1)$$

That is V_n is covered by F-sets centered on the elements of B.

Proposition 2 . If a block B is good, any translate B+t, $t \in V_n$, of B is also good.

Proof :
$$\bigcup_{b' \in B+t} F(b') = \Big(\bigcup_{b \in B} F(b)\Big) + t = V_n . \ \square$$

Proposition 3. If B is good, then :

$$\forall x \in V_n \quad \exists b \in B \quad \exists f \in F : x + f = b.$$

In other words, starting from any state x of the memory, there exists an allowed transition f which transforms x into an element of B (say b).

Proof. By (1), for all x, there is an i s.t. x is in $F(b_i)$, i.e. $x = b_i + f$ for some f in F. □

Proposition 4. If $B_0, B_1, \ldots B_{M-1}$, are pairwise disjoint good blocks, they yield a W∗M-code of size M.

Proof. Put the M messages to be coded in 1-1 correspondence with the blocks. By Proposition 3, whatever the state of the W∗M is, updating will be possible to any message. □

<u>Example for a WIM</u>. $n = 3$. Set $B_o = \{000,111\}$. Then B_o is good, since :

$$F(000) = F = \{000, 001, 010, 100, 101\}$$

$$F(111) = \{111, 110, 101, 011, 010\} = \overline{F}$$

and $F \cup \overline{F} = V_3$, so (1) holds.

By Proposition 2, the following blocks are also good :

$B_1 = B_o + 001 = \{001, 110\}$; $B_2 = B_o + 010 = \{010, 101\}$,
$B_3 = B_o + 100 = \{100, 011\}$.

This yields by Proposition 4, since B_o, B_1, B_2, B_3 form a partition of V_3, a WIM-code with 4 codewords, i.e. rate 2/3. Unfortunately, we could not use this example for an infinite construction.

Let us visualize how the coding works : suppose we are in state $x = 010$, representing message 2. The following are the allowed transitions for writing messages 0,1,2,3 respectively :

010 --> 000, 010 --> 110, 010 --> 010, 010 --> 011.

We shall now show that the upperbound in Proposition 1 is tight. This will a fortiori give the capacity in the more favorable case 1 when writer and reader know the previous state. This result is not difficult to prove in a probabilistic (non constructive) way. We shall rather give here a "semi-constructive" proof, which also helps in obtaining good codes.

In view of Proposition 4, it is intuitive to look for "small" good blocks, so as to be able to pack many of them (e.g. by translation) in

In fact, we shall first prove the existence of small <u>good subgroups</u> of V_n (i.e. good blocks which are groups). Then the second step, finding pairwise disjoint good blocks, becomes simple : if G is a good subgroup, $|G| = 2^k$, then there are 2^{n-k} pairwise disjoint good blocks, namely the cosets of G (see example 2). To that end, we use Theorem 1 of [4], which is established for coverings of V_n by Hamming spheres centered on the elements of a group (group coverings). Its extension to group covering by tiles other than spheres is easy and already mentioned in [4], so we shall not give its proof, based on a "group" greedy algorithm :

Proposition 5. There exists a group covering G of V_n with 2^k sets $F(g_i)$, $g_i \in G$, with

$$k \leq n - \log f_n + \log n + 0(1). \square$$

This gives :
$M = 2^{n-k} \geq f_n / n.0(1)$, and the following result.

Proposition 6. $$C_{0,1} = C_{0,2} = \lim_{n\to\infty} n^{-1} \log f_n . \square$$

Dropping the group condition, one can obtain still smaller good blocks, but this will of course not improve the rate. We shall nevertheless give some details, since they shed more light on possible constructions.

Let $B = \{b_1, b_2, \ldots b_m\}$ be a good block of minimal size (i.e. by (1) a minimal covering of V_n by F-sets). Consider the hypergraph $H = (V,E)$, where $V = V_n$ and $E = \{F(x), x \in V_n\}$.
Then H is clearly f_n - uniform and f_n-regular (i.e. $|\{x : y \in F(x)\}| = f_n$ for all y). Thus by a Theorem of Lovàsz [15], there exists a covering with

$$2^n/f_n \leq m \leq (2^n/f_n)(1 + \log f_n),$$

where the lower bound is the well-known covering bound.

5. THE ε-ERROR CASE

We now deal with the following simple probabilistic model : every position of the memory is defective, reluctant,.. with some probability p. All associated random variables are assumed i.i.d. Hence the number w of these positions is a random variable with expectation $E(w) = np$ and variance $np(1-p)$. The 0-error capacity is clearly 0 here, and the relevant quantity is the ε-error capacity, i.e. the maximal rate achievable with vanishing probability or error.

This is the quantity commonly used in information theory under the name of (Shannon) capacity.

We shall focus on 3 problems of increasing difficulty :

writing on defective, suspect and erroneous positions (WDM, WSM and WERM respectively).

1. WDM. We only consider case 2 (encoder informed, decoder non informed). The capacity is [10]

$$C_{\varepsilon,2} = 1 - p.$$

Note that when exactly s defects occur, we have already seen in § 3 and § 4 that $C_{0,2} = 1 - sn^{-1}$.

Easy modifications of the semi-constructive greedy algorithm given in § 4 yield codes achieving $\mathbf{C}_{\varepsilon,2}$.

2. WSM. This problem is introduced in [2] under the name : "Coding for channels with localized errors" : the encoder knows a set S(with E(|S|=s=pn) of suspicious positions, i.e. positions in which an error can occur with probability 1/2 when writing. Again, we are in case 2. The authors prove

$$C_{\varepsilon,2} = 1 - H(p).$$

This is disappointing in one sense, because this quantity is smaller than for WDM and in fact equals the capacity for errors (see next paragraph) ! A nonintuitive fact since the encoder knows definitely more than when errors occur. On the other hand this side-information facilitates the encoding. To see this, consider the following coding strategy :

- encode the positions of the suspicious cells (this requires $\log\binom{n}{pn} \cong n\,H(p)$ bits)

- code information on the remaining cells.

This strategy achieves $R = 1 - H(p) - p$.

3. WERM.

This is the fondamental problem of coding theory. We are in case 4. The memory behaves like a binary symmetric channel (with transmission in space rather than in time). It is well known that

$$C_{\varepsilon,4} = 1 - H(p).$$

No effective coding procedure approching capacity is available.

W*M

TABLE OF $C_{i,j}$ and $R_{i,j}$ ($i = 0,\varepsilon$ and $j = 1, 2, 3, 4$)

		0-error cases 1	2	3	4	ε-error cases 1	2	3	4
Encoder informed		yes	yes	no	no	yes	yes	no	no
Decoder informed		yes	no	yes	no	yes	no	yes	no
WUM	R	Θ	0.563[(1)]	0.528	0.5	Θ			
	C	Θ	Θ	0.528?	0.5?	Θ	Θ	Θ[(2)]	0.545[(3)]
WIM	R	Θ	0.6[(4)]	Θ	0.5	Θ		Θ[(5)]	
	C	Θ	Θ	Θ	0.5?	Θ	Θ	Θ	0.5?
WRM	R								
	C		H(p)						
WDM	R								
	C		1-p				1-p		
WSM	R						1-H(p)-p		
	C						1-H(p)		
WERM	R								
	C								1-H(p)

$\Theta = \log((1+\sqrt{5})/2) \cong 0.694$

? is a conjectured value

(1) see Koschnick

(2) van Overveld

(3) Godlewski, Wyner, Ozarow, Willems

(4) Cohen-Zemor

(5) see following discussion

Comments on the table

R denotes the rate achieved by constructive methods, i.e. with algorithms whose size is a polynomial in the size of the inputs. Hence the methods in §4 are not constructive, though explicit.

For a WIM, $C_{0,3}$ is obtained in the following way :

The $2^{\gamma n}$ Fibonacci sequences (where $\gamma = (1 + \sqrt{}5/2)$ is the Golden ratio) of the set

$$F(\underline{0}) = \{(x_1, x_2, \ldots, x_n) \in V_n : x_i x_{i+1} = 0 \text{ for } i = 0, 1, \ldots, n-1\}$$

can be numbered using the Fibonacci number system (see, e.g. [19]).

The inverse operation, namely

$$j \in [1, 2^{\gamma n}] \rightarrow f(j) \in F(\underline{0})$$

is also feasible in polynomial time. Hence the coding and decoding procedures are the following :

CODING OF j $\quad 1 \leq j \leq |F|$

1. Compute f(j)

2. Add f(j) to the previous state x(t) (this needs no reading, only complementing some positions), getting x(t+1).

DECODING

1. Set $x(t+1) + x(t) = f \in F(\underline{0})$
   (This is possible since the decoder knows x(t) and x(t+1))

2. Convert f into the number j associated to it.

REFERENCES

[1] R. AHLSWEDE, Z. ZHANG, *Coding for write-efficient memories,* Inform. and Control, vol. 83, n°1 (1989) 80-97.

[2] L.BASSALYGO, S.GELFAND and M.PINSKER, *Coding for channels with localized errors,* Oberwolfach Tagungsbericht 21/1989, May 1989.

[3] J.M. BORDEN, *Coding for write-unidirectional memories, preprint.*

[4] G. COHEN, P. FRANKL, *Good coverings of Hamming spaces with spheres,* Discrete Math. 56 (1985) 125-131.

[5] G. COHEN, P. GODLEWSKI and F. MERKX, *Linear block codes for write-once memories,* IEEE Trans. Inform. Theory, IT-32, n°5 (1986) 697-700.

[6] G. COHEN, G. SIMONYI, *Coding for write-unidirectional memories and conflict resolution,* Discrete Applied Math. 24 (1989) 103-114.

[7] G. COHEN, G. ZEMOR, *Write-Isolated Memories,* French-Israeli Conference on combinatorics and algorithms, Nov. 1988, Jerusalem, to appear in Discrete Math.

[8] I.CSISZAR and J.KORNER, Information Theory, Academic Press

[9] M.R. FELLOWS, *Encoding graphs in graphs,* Ph.D. Dissertation, Univ-Calif. San Diego, Computer Science, 1985

[10] C.HEEGARD and A.A. EL GAMAL, *On the capacity of Computer Memory with Defects,* IEEE Trans. on Inform. Theory, vol. IT-29, n°5, (1983) 731-739.

[11] T. KLØVE, *On Robinson's coding problem,* IEEE Trans. on Inform. Theory, IT-29, n°3 (1983) 450-454.

[12] K.U. KOSCHNICK, *Coding for Write-Unidirectional Memories*, Oberwolfach Tagungsbericht 21/1989, May 1989.

[13] A.V. KUZNETSOV, *Defective channels and defective memories*, Oberwolfach Tagungsbericht 21/1989, May 1989.

[14] A.V. KUZNETSOV and B.S. TSYBAKOV, *Coding in memories with defective cells*, Probl. Peredachi, Inform., vol. 10, n°2 (1974) 52-60.

[15] L. LOVASZ, *On the ratio of optimal integral and fractional covers*, Discrete Math. 13 (1975) 383-390.

[16] F.J. MACWILLIAMS and N.J.A. SLOANE, The Theory of Error-correcting Codes, North-Holland, New-York, 1977.

[17] W. M.C.J. van OVERVELD, *The four cases of WUM-codes over arbitrary alphabets*, submitted to IEEE Trans.on Inform. Theory.

[18] R.L. RIVEST, A. SHAMIR, *How to reuse a "write-once" memory*, Inform. and Control 55 (1982) 1-19.

[19] M.R. SCHROEDER, Number Theory in Science and Communication, Springer-Verlag Series in Information Sciences, 1984.

[20] G. SIMONYI, On Write-Unidirectional Memory Codes, IEEE Trans. on Inform. Theory, vol. 35, n°3 (1989) 663-669.

[21] A. VINCK, *personal communication.*

[22] H.S. WITSENHAUSEN and A.D. WYNER, On Storage Media with Aftereffects, Inform. and Control, 56 (1983), 199-211.

[23] J.K. WOLF, A.D. WYNER, J. ZIV, J. KÖRNER, *Coding for write-once memory*, AT and T Bell Lab.-Tech. J. 63, N°6 (1984) 1089-112.